T0269464

SpringerBriefs in Philosophy

Philosophy of Science

SpringerBriefs in Philosophy of Science will consist of original works in the philosophy of science that present new research results or techniques that are of broad instructional significance at the graduate and advanced undergraduate levels.

Topics covered will include (but not be limited to):

Formal epistemology
Cognitive foundations of science
Metaphysics of models, laws, and theories
Philosophy of biology
Philosophy of chemistry
Philosophy of mathematics
Philosophy of physics
Philosophy of psychology
Philosophy of the social sciences

The series is intended to bridge the gap between journal articles and books and monographs. Manuscripts that are too long for journals but either too specialized or too short for books will find their natural home in this series. These will include suitably edited versions of lectures and workshop presentations that develop original perspectives on the philosophy of science that merit wide circulation because of the novelty of approach.

The length of each volume will be 75–125 published pages.

More information about this series at http://www.springer.com/series/13349

Prasanta S. Bandyopadhyay
Gordon Brittan Jr. · Mark L. Taper

Belief, Evidence, and Uncertainty

Problems of Epistemic Inference

 Springer

Prasanta S. Bandyopadhyay
Montana State University
Bozeman, MT
USA

Mark L. Taper
Montana State University
Bozeman, MT
USA

Gordon Brittan Jr.
Montana State University
Bozeman, MT
USA

ISSN 2211-4548 ISSN 2211-4556 (electronic)
SpringerBriefs in Philosophy
ISSN 2366-4495 ISSN 2366-4509 (electronic)
Philosophy of Science
ISBN 978-3-319-27770-7 ISBN 978-3-319-27772-1 (eBook)
DOI 10.1007/978-3-319-27772-1

Library of Congress Control Number: 2016930269

This Springer imprint is published by SpringerNature
The registered company is Springer International Publishing AG Switzerland

To my mother, Gouri Devi

Prasanta S. Bandyopadhyay

*To the memory of my parents, Gordon
and Thelma Brittan*

Gordon Brittan Jr.

*To my late mother Phyllis A. Whetstone Taper,
my father Bernard B. Taper, my wife
Ann P. Thompson, and my
daughter Oona A. Taper*

Mark L. Taper

Preface

> *If you want to go fast, go alone.*
> *If you want to go far, go together.*
>
> African Proverb

This monograph has two main aims: to make precise a distinction between the concepts of confirmation and evidence and to argue that failure to make it, in our terms, is the source of certain otherwise intractable epistemological problems. Confirmation has to do with the adjustment of our beliefs in light of the data we accumulate, and is therefore "in the head" and in this sense "subjective;" evidence, however, has to do with the relationship between hypotheses or models and data, accumulated or not, and is "in the world" or "objective." A subsidiary aim is to demonstrate to philosophers the fundamental importance of probabilistic and statistical methods not simply to inferential practice in the various sciences, where they are now standard, but to epistemic inference in other contexts as well.

The argument for the main aim depends in turn on three others:

1. That the best-known attempts to characterize a satisfactory concept of evidence in terms of Bayesian or other non-Bayesian confirmation theories fail;
2. That the standard error-statistical, "bootstrap," and "veridical evidence" accounts of hypothesis testing are not as successful as the one developed in this monograph;
3. That some traditional epistemological puzzles are solved in a clear and straightforward way once the confirmation/evidence distinction is made along our lines.

Although the argument is rigorous, it is also accessible. No technical knowledge beyond the rudiments of probability theory, arithmetic, and algebra is presupposed, symbols are kept to a minimum, otherwise unfamiliar terms are always defined, and a number of concrete examples are given. More specialized material has been placed in Appendices to several of the chapters. That our line of argument is at least initially somewhat counterintuitive should make it all the more interesting and important to philosophers and the philosophically-minded.

The first part of the African proverb above can be taken to promote single authorship of a monograph for expeditious publication. It could have been written much more quickly by any one of us. But "going it alone," we would not have been able to go as far as we have collectively. We each brought ideas from our varying specializations and created a synergy that propelled the work to what we hope are far-reaching ends. We envision that it will spur subsequent discussion of statistical and epistemic reasoning by philosophers, as well as their consideration by scientists interested in a larger view of their own inferential techniques.

Acknowledgments

The materials for this monograph have been assembled over a long period of time. In the process, we have very much benefitted from the numerous and very helpful comments of Marshall Abrams, Robert Boik, Earl Conee, David Cox, David Christensen, Abhijit Dasgupta, Brian Dennis, Yonaton Fishman, Malcolm Forster, Andrew Gelman, Jack Gilchrist, Peter Gerdes, Clark Glymour, Mark Greenwood, James Hawthorne, Colin Howson, James Joyce, Subhash Lele, Adam Morton, José Ponciano, Jan-Willem Romeijn, Glenn Ross, Tasneem Sattar, Elliott Sober, Teddy Seidenfeld, and Paul Weirich. Thanks are also due Michael Barfield, whose keen eye for detail and perfection made of this manuscript a much better text. Special thanks are due to John G. Bennett for years of conversations, both e-mail and in person, regarding the issues discussed.

The support of Hironmay Banerjee, David Cherry, Birdie Hall, Chhanda Gupta, Michelle Maskiell, and Brian Skyrms during the preparation of this manuscript is very much appreciated, as is the technical help of our superb administrative staff, Kori Robbins and Kathryn Yaw.

We would also like to thank the Springer-Verlag referees and the series Editorial Board for their very helpful comments, and the series Editor, Sahotra Sarkar, for his suggestions regarding placing our research program in the context of current philosophy of science.

We very much appreciate the help we received from the Springer staffs; Banu Dhayalan, Christi Lue, and Werner Hermens regarding meeting different deadlines, and quality production of the book from the very beginning to the end.

This work is a collaboration not simply in the sense that it is the product of our joint efforts, but perhaps more importantly because it bridges disciplines and conjoins very different points of view. Prasanta Bandyopadhyay and Gordon Brittan are philosophers whose orientation with respect to uncertain inference is generally Bayesian. Mark Taper is a statistical ecologist for whom Bayesian concepts play no role in the day-to-day analysis of data and the appraisal of scientific hypotheses, but for whom beliefs do play an important role in the establishment of research programs. The basic theme of the monograph, that a theory of uncertain, and for that reason inductive, inference must incorporate a fundamental distinction between justification or confirmation and evidence, expresses these differences even as it attempts to make room for both of them. In our view, it is very discouraging that proponents of one statistical paradigm or another continue to confront each other in an adversarial, occasionally even bitter, way. That scientists working on a very wide variety of problems which require probabilistic techniques are making great strides forward should prompt deeper reflection on the part of philosophers, statisticians, and scientists concerning our respective agendas.

Contents

Part I
Clarification, Illustration, and Defense of the Confirmation/Evidence Distinction

Chapter 1
Introductory Remarks—Inferential Uncertainty

Abstract Most of the claims we make, nowhere more so than in the empirical sciences, outrun the information enlisted to support them. Such claims are never more than probable/likely. Intuitively, even obviously, some claims are more probable/likely than others. Everyone agrees that scientific claims in particular are probable/likely to the extent that they are confirmed by experimental evidence. But there is very little agreement about what "confirmation by empirical evidence" involves or how it is to be measured. A central thesis of this monograph is that a source of this disagreement is the near-universal tendency to conflate the two different concepts—"confirmation" and "evidence"—used to formulate the essence of the methodology. There is no doubt that the words signifying them are used interchangeably. But as we will go on to argue, failure to make the distinction leads to muddled thinking in philosophy, statistics, and the empirical sciences themselves. Two examples, one having to do with the testing of traditional psychotherapeutic hypotheses, the other with determining the empirical superiority of the wave or particle theories of light, make it clear how data can confirm a hypothesis or theory, yet fail, in context, to provide evidence for it.

Keywords Belief · Evidence · Inferential uncertainty · Dodo bird verdict · Poisson spot

The "Experimental Method", Belief, and Evidence

We begin with a paradox. At least since the time of Galileo, "science" has virtually been defined in terms of its use of the "experimental method", to wit, the tenet that its various hypotheses must be confirmed by empirical evidence if they are to be accepted. This is the widely-acknowledged source of the pre-eminent credibility that many scientific theories have enjoyed since the 16th and 17th centuries. Yet there is as yet very little agreement among philosophers and philosophically-minded statisticians about what such "confirmation by evidence" amounts to, or how it is to be analyzed and understood. Indeed, the situation is even worse. Without exception,

© The Author(s) 2016
P.S. Bandyopadhyay et al., *Belief, Evidence, and Uncertainty*,
Philosophy of Science, DOI 10.1007/978-3-319-27772-1_1

the standard philosophical analyses advanced have led to paradoxes, and the dis-agreements among those involved seem to grow greater by the day. We need to set out on a new path.

Among much else, Immanuel Kant left philosophers two pieces of superb methodological guidance. One is contained in the opening words of the second section of *Thoughts on the True Estimation of Living Forces*:

> … An insight that I have always used as a rule of the examination of truths. If men of good sense, who either do not deserve the suspicion of ulterior motives at all, or who deserve it equally, maintain diametrically opposed opinions, then it accords with the logic of prob-ability to focus one's attention especially on a certain intermediate claim that agrees to an extent with both parties.[1]

This might be taken to mean: if men of good sense maintain diametrically opposed opinions, then it is a wise move to split the difference between them, establishing some sort of common ground. Indeed, Kant's practice often suggests as much, that the truth generally lies between extremes; both rationalism and empiricism, for example, have something to be said in their favor, and the best strategy is to begin at those places where they intersect.

We take Kant's advice in another way, however. On our reading, when neither of two sides is willing to relent, and more especially when it seems that the arguments on each side are cogent, we should look for premises that both accept, and examine them more closely. On this reading, whatever common ground the two sides to an apparently endless dispute share is itself suspect, a likely source of their ultimate disagreements.

The other piece of methodological advice Kant left us confirms our way with the first. It has two aspects. On the one hand, philosophy proceeds largely, if not entirely, by way of resolving paradoxes or what he called "antinomies" in thinking about the world and our knowledge of it. On the other hand, such resolution is obtained by making crucial distinctions. In the case of the debate between Cartesians and Leibnizians which is the focus of *Thoughts on the True Estimation of Living Forces*, the distinction to be made is between two very plausible senses of "living force", between what we now term "momentum", mv, and "mean kinetic energy", $\frac{1}{2}$ mv^2. The heretofore shared premise, that there is but one plausible sense, must be given up and the dispute is resolved.

Our two-fold argument in this monograph is straightforward. It is that philosophers who otherwise disagree about the character of scientific and other kinds of epistemic inference share a common assumption, that the concepts of evidence and confirmation are but two sides of the same coin, and that the holding of this assumption is the source of many of the seemingly intractable, often para-doxical, difficulties into which they fall.

In the process of making this distinction as precise as possible, confirmation in terms of up-dating probabilities by way of Bayes Theorem and evidence in terms of

---

[1]Section II, #20.

the ratio between likelihoods,[2] we will draw on a wide array of probabilistic tools and statistical methods. We hope to demonstrate to those not already convinced that these tools and methods provide indispensable help in the clarification and eventual solution of questions surrounding the character of scientific inference in particular and the grounding of human beliefs in general. It is a source of continuing concern to us that while multiple courses in logic and set theory continue to be part of the philosophy curriculum everywhere, comparatively little instruction in probability theory and statistics is offered. This is somewhat puzzling in light of the fact that David Hume, Bishop Butler, and other philosophers in the 18th century focused so much attention on probabilistic concepts,[3] as did Hans Reichenbach, Rudolf Carnap, and other leading philosophers of science in the 20th century.[4]

Inferential Uncertainty

Most of the claims we make, nowhere more so than in the empirical sciences, outrun the information enlisted to support them. This can be brought out in two different and familiar ways. On the one hand, however much information has been gathered, it is still possible that the claims they are taken to support are false. As Hume famously said, for all we know, the sun might not rise tomorrow morning. The conclusions of such arguments are never more than merely probable/likely. On the other hand, adding information to the premises of deductive arguments does not affect their validity. If p & q logically entails p, so too does p & q & r. But if a foxy scent in the air supports the claim that we are on the track of a fox, it does not so soon as we add the information that this is a "dry hunt" (i.e., someone has gone ahead to lay down a scent for the hounds to follow). In this case (of non-monotonic reasoning), the claims we make are hostage to the information taken to support them and in this sense as well inevitably outrun that information. Once again, such claims are never more than probable/likely. Intuitively, even obviously, some such claims are more probable/likely than others. It is the task of uncertain inference to say under what conditions and to what extent this is so. It is customary in the first sort of case, when it is possible for the informational premises to be true and the claim made on their basis false, to talk about "inductive probability", in the second sort of case, when the credibility of a claim rests on the kind of evidence available to support it, to talk about "epistemic probability", the one attaching to arguments and the other to statements, but it should be clear that the probability of

---

[2]The likelihood ratio is one among a number of evidence functions. We will later list the reasons for its choice.

[3]See Hacking (1975), for a lucid account of the history.

[4]Perhaps it can be explained in terms of the fact that philosophical claims are traditionally taken as true or false, to be established or refuted on the basis of deductive arguments. Logic and set theory are well suited to appraise their validity. In science, on the other hand, there is a much greater emphasis on quantitative models which are all strictly false, but whose adequacy is best assessed probabilistically.

statements is tied closely to their figuring as conclusions of inductive arguments. We might bring these considerations together under the heading of formal or *inferential uncertainty*. It has been a source of philosophical concern at least since the time of Hume.[5]

There is more. It is not simply that many of the claims we make are sensitive to and invariably outrun the information we have in hand, but that we live in an increasingly complex world, whose description and *a fortiori* its adequate explanation(s) requires probabilistic and statistical concepts. We might bring this fact under the heading of *substantive uncertainty*, by which we understand statistical as well as quantum uncertainty. It has been acknowledged as such at least since Bernoulli and Pascal in the 17th century and Heisenberg in the 20th.[6]

Our concern in this monograph is with inferential uncertainty.[7] To re-state our initial paradox in more contemporary terms, our ability to explain and describe the world goes far beyond whatever abilities to do so our ancestors might have had, yet the inferential structures which support this ability of ours are the source of continuing controversy and their own uncertainty. Still, we have some very elegant and well-developed probabilistic and statistical tools with which to work. Our aim here is not to extend but to re-deploy them, in the expectation that at least some of the controversy, if not also uncertainty, will be mitigated.

The Action Agenda

In the final section of this chapter, we will document the way in which the notions of confirmation and evidence have been conflated and sketch two examples that indicate why and how they should be distinguished. In Chap. 2, the distinction between the two notions is made formally precise, and in Chap. 3, several examples

---

[5]Hume understands "By probability, that evidence, which is still attended with uncertainty. 'Tis this last species of reasoning I propose to examine" (1888/1739–1740, p. 124). It is significant that in this paragraph from *A Treatise of Human Nature*, Hume attributes "degrees" to evidence, associates it with "reasoning" or inference, and (implicitly) distinguishes it from "experience" (by which we understand "data"). In all three respects, our discussion follows his and breaks with what has come to be the conventional wisdom.

[6]One can distinguish between statistical and quantum uncertainty, but not isolate them. A certain amount of statistical uncertainty, on occasion not insignificant, will be due to quantum uncertainty. Butterfly-like amplifications translate some quantum uncertainty into statistical uncertainty. The behavior of animals responding to chemoreceptors is an example. Some chemoreceptors are so exquisitely sensitive that they will respond to single molecules. Whether or not a particular molecule comes into contact with a receptor depends on the quantum uncertainty of Brownian motion.

[7]The distinction between inferential and substantive uncertainty, the one having to do with the kinds of conclusions we humans are capable of drawing, the other with the course of natural phenomena, is not sharp. Statistical distributions are natural phenomena, but they often give the arguments in which they figure as premises an inductive character.

are given to illustrate it further and demonstrate its importance. Chapter 4 replies to several of the most prominent difficulties that have been raised in connection with our way of formulating the confirmation/evidence distinction, and in the process clarifies our intentions.

The first part of the monograph thus sets out, develops, and defends our distinction between evidence and confirmation. The second part of the book discusses alternative accounts of hypothesis-testing which either dispense with the distinction or cast it in different terms. Chapter 5 considers, and then rejects, a well-known attempt by David Christensen, James Joyce, and Alan Hájek to make something like the distinction, or least to provide an adequate surrogate for "evidence", in Bayesian "confirmational," terms alone. Chapter 6 takes up Deborah Mayo's well-known error-statistical ("Neyman–Pearson"-based) account of the severe testing of hypotheses, develops some key difficulties with it, and notes the advantages of our own. Clark Glymour both rejects the Bayesian (and more generally probabilistic) approach to uncertain inference and adds influential "bootstrap" conditions to some of its traditional alternatives. In Chap. 7, we examine his account, find problems in it stemming from its conflation of evidence and confirmation, and as a corollary show in an intuitive way how the problem of selective hypothesis testing can be solved. Peter Achinstein has developed another very interesting quasi-probabilistic conception of evidence which, we argue in Chap. 8, similarly confounds evidence with confirmation and, in addition to other problems, eventually runs afoul of some powerful theorems about the probability of misleading evidence.

The third part of the monograph makes a case for the usefulness of the distinction between confirmation and evidence by redeploying it to solve some epistemological puzzles. Chapter 9 sets out and then resolves three so-called "paradoxes of confirmation"—the Raven, Grue, and Old Evidence Paradoxes—in terms of it. Chapter 10 discusses Descartes' celebrated argument for radical doubt about the existence of a mind-independent world, the argument from dreaming, and attempts to show both how it depends on a conflation of evidence and justification and instantiates a broader pattern of arguments for paradoxical and skeptical claims. Finally, in Chap. 11, we put our argument in a larger perspective and point out a new direction in understanding uncertain and epistemic inference. Relatively brief, somewhat technical, appendices which amplify and make more precise certain crucial issues discussed in the body of the monograph have been attached to Chaps. 2, 6, and 11. The Appendix to Chap. 11 addresses several of these issues and, although mathematically sophisticated, should serve to put them at rest.

Confirmation and Evidence Conflated

Open any one of the numerous and excellent anthologies of work in epistemology. In their introductory commentaries, the notions of confirmation (used interchangeably with "justification") and evidence are invariably run together. Thus

(Bernecker and Dretske 2000, p. 3): "what distinguishes knowledge from mere true belief and lucky guessing is that it is based on some form of justification, evidence, or supporting reasons". Or again, (Sosa et al. 2000, p. 190): "a justified belief is one in which the available evidence makes the truth of the belief very likely". Even when the context is not introductory and casual, the same assimilation between evidence and confirmation or justification is made. (Bonjour 2002, p. 41), for example, asserts without further ado that "the concepts of an epistemic reason or of an epistemic justification as they figure in the traditional concept of knowledge are, if not simply identical to the concept of evidence, at least fairly straightforward generalizations of that concept". As yet one more example, (Kim 1988) insists on their assimilation: "…the concept of evidence is inseparable from that of justification. When we talk of 'evidence' in an epistemological sense we are talking about justification; one thing is 'evidence' for another just in case the first tends to enhance the reasonableness or justification of the second".

There are many different ways in which contemporary epistemological theories are classified. They are variously internalist or externalist, foundationalist or coherentist, naturalized or normative. But almost all of them seem to share, over and above a common interest in answering epistemic questions, the assumption that "evidence" and "confirmation" or "justification" are congruent notions.[8] This assimilation of evidence and confirmation is perhaps most striking in conventional Bayesian epistemology, where data D are usually taken as evidence for hypothesis H just in case $\Pr(H \mid D) > \Pr(H)$, that is to say, just in case our degree of belief in H given D is greater than our degree of belief in, or confirms, H.

The situation is no different with respect to the philosophy of science, where the focus is more narrowly on scientific inference.[9] Evidence and confirmation are habitually run together. Carl Hempel, for a notable example, writes in his classic

---

[8]There is little point in multiplying examples from the standard literature. Suffice it to note that advocates of "reliabilism," perhaps the currently most popular "naturalized" epistemological theory, assimilate evidence and justification without apology. See Alvin Goldman's seminal "A Causal Theory of Knowing" (Goldman 1967); he equates "a highly warranted inductive inference" with one which gives an epistemic agent "adequate evidence" for his belief. The most recent comprehensive summary of theories of confirmation with which we are familiar, the article "Confirmation" in the *Stanford Encyclopedia of Philosophy* (Crupi 2013/2014) simply assumes that evidence and confirmation are inter-definable regardless of the confirmation theory at issue, which is to say that no attempt is made by any of the theories discussed to distinguish "data" from "evidence". For a more popular on-line assimilation, see the helpful article entitled "Evidence" in the *Internet Encyclopedia of Philosophy* (DiFate 2007): "In the philosophy of science, evidence is taken to be what confirms or refutes scientific theories" (p. 2) and again "evidence is that which justifies a person's belief" (p. 7), although this latter claim is not simply assumed but argued for. It is worth adding, as a note to those who have told us on occasion that the discussion of evidence has advanced well beyond our characterization of it, that this relatively recent summary of the current situation focuses on exactly those positions and people that we do here.

[9]Not simply the inference at stake when we infer the description of and thus predict future events, but identify and select models to describe natural processes, estimate and select the values of parameters in the descriptive models, and assess the consistency of the evidence with all three. See, for example, (Cox 2006).

paper, "Studies in the Logic of Confirmation", that "an empirical finding is relevant for a hypothesis if and only if it constitutes either favorable or unfavorable evidence for it; in other words, if it either confirms or disconfirms the hypothesis".[10] The "in other words" indicates that for Hempel, the connection between evidence and confirmation is self-evident. As we shall see in more detail in what follows, almost every succeeding philosopher of science echoes something like this conflation.[11]

There is little point in denying that the words "evidence" and "confirmation" (or in the present context, "justification") are often used interchangeably, in ordinary English as well as among scientists, philosophers of science and inductive logicians.[12] Our case for distinguishing two concepts that the words can be used to identify rests not on usage, but on the clarification in our thinking that is thus achieved. Making the distinction between concepts allows us to solve some long-standing paradoxes and problems. The main task of some of the chapters to follow will be to spell these solutions out in detail. But it might help here to consider another case where advancing an initially counter-intuitive distinction contributed in important ways to philosophical progress.

Before 1905, the sentence "the present King of France is bald" was universally taken as subject-predicate in form, and the definite description "the present King of France" within it as a singular term.[13] But in the 20th century's most celebrated English-language philosophy paper, "On Denoting", Bertrand Russell argued that this was simply an illusion, fostered by surface grammar, which did not in any way correspond to the deep structure reality. "The present King of France is bald" is to be rendered correctly as "$(Ex)\{[Kx \& (y)Ky \rightarrow (x = y)] \& Bx\}$", i.e., "there is an object x such that it is King of France, and for all objects y, if any one of them is King of France it is identical to x, and x is bald". Intuition to the contrary, "The present King of France is bald" is not really of subject-predicate form and "the present King of

---

[10]Hempel (1965, p. 5) Rudolf Carnap, too, assimilates evidence and confirmation in underwriting the idea that degree of belief should be identical to weight of evidence. See his "Statistical and Inductive Probability," reprinted in Brody (1970, p. 441).

[11]There are many good features of Bovens and Hartmann's highly original application of both probability and Bayesian-network theories to solving philosophical problems (Bovens and Hartmann 2004). Our approach differs from theirs, however, in three respects. First, unlike their "engineering approach," ours is a foundational investigation into issues concerning "evidence" and "confirmation". Second, and again unlike theirs, ours is not simply an exercise in "Bayesian epistemology". Third, like almost all writers on confirmation theory, they fall victim to the conflation of evidence and confirmation (see especially their chapter on confirmation).

[12]Or in denying that "evidence" and "confirmation" has each been endowed with a variety of different meanings. It is entirely typical of the literature, however, that in a foundational article, "Confirmation and Relevance" (Salmon 1975), Wesley Salmon first notes the ambiguities and then proceeds to focus on "investigating the conclusions that can be drawn from the knowledge that this or that evidence confirms this or that hypothesis". Even, and perhaps especially, the very best philosophers working in the area cannot resist conflating the two concepts.

[13]At least this is the story we have long learned to tell. It is questioned, to put it mildly, by Oliver (1999).

France" is not a term. Neither ordinary language nor philosophical tradition informs us of this fact. Rather, Russell argues for his analysis on the basis of another fact, that it solves a variety of notorious puzzles, the problem of attributing truth-values to sentences in which the apparent subject-term fails to refer among them. In the process, he made a distinction between logical and grammatical form that few people, not even Frege (at least not with respect to terms) had ever made, and set analytic philosophy a central task, to make the former perspicuous.

"Evidence" and "confirmation" (or "justification") are often used interchangeably. This is certainly clear, even from the comparatively few passages cited. But we also believe that this conflation, whatever terms are used to denote two otherwise distinct concepts, is the source of a great deal of error in philosophy, statistics, and the practice of science. In what follows, we will try to make good on this claim so far as philosophy in particular is concerned. In the next chapter, the distinction we have in mind will be made precise. For the moment, two well-known examples, one from applied psychology, the other from theoretical physics, should help make our fundamental insight more intuitive, and, we hope, motivate its further elaboration, discussion, and application.

In 1936, Rosenzweig published a ground-breaking paper, "Some Implicit Factors in Diverse Methods of Psychotherapy".[14] In it, he proposed that the features common to different therapeutic methods, among them the therapist's personal interest in the patient's recovery, led to generally positive outcomes for all of them. In this connection, he quoted the Dodo-bird's verdict in *Alice in Wonderland* after judging a race: "everybody has won so all shall have prizes". What Rosenzweig supposed, that in virtue of their common factors the various psychotherapies were to a large degree equally efficacious, has thus come to be called "The Dodo-bird Verdict". Whatever the form of psychotherapy in practice, more patients feel better after treatment than before, their psychological-behavioral symptoms are to one extent or another relieved. Given the greater number of positive outcomes, it is natural to conclude that in general the psychotherapies "work", that is, that the efficacy of each has been confirmed. This is certainly what many therapists claim: the fact that a particular method—Freudian, Jungian, Rogerian, whatever—raises the probability of a patient's symptomatic relief demonstrates the validity of the method.

An enormous literature has sprung up in the wake of Rosenzweig's paper. Although the "Dodo-bird Verdict" remains controversial in some quarters, most subsequent research seems to support it.[15] That is, the diverse psychotherapies still in use tend to be more or less equally efficacious, the factors common to them all

---

[14]Rosenzweig (1936).

[15]See Luborsky et al., (2002) for an especially rigorous examination of 17 meta-analyses of the relevant data. What follows in the above paragraph draws on this examination. The article includes an extensive bibliography.

explain the beneficial results, and, it comes as no surprise, most of the studies that appear to show that one psychotherapeutic method is clearly superior to others have been carried out by researchers who were in one way or another already biased toward that method.

But the fact that all of the psychotherapies studied "win", and in this sense are confirmed, is unsettling, for the methods employed by one or the other vary, in precisely those respects underlined by their practitioners, radically. What we want to say is that while all are "confirmed", there is no evidence that one is superior to the others. "Evidence" for all is evidence for none; a race where everyone wins has no winner. To say this, in turn, is to say that the likelihood of positive outcomes on one or the other treatment is roughly the same.[16] The available data do not favor one over the other.[17]

The implication is that what we term "confirmation" should be distinguished from "evidence". Confirmation is agent-dependent in the sense that a hypothesis is confirmed just in case an agent's degree of belief in it is raised. It is in just this sense that the diverse psychotherapies have been confirmed or justified by and for their practitioners and patients. It is in this same sense in part subjective, starting with the element of belief. Evidence, however, is agent-independent; it has to do not with raising prior degrees of belief in a hypothesis on the basis of the data collected, but in assessing the relative likelihood of the data given one or the other of the two hypotheses. It is in this sense "objective".[18] It is also comparative. Evidence consists of data more likely on one hypothesis than another. The greater the likelihood ratio, the stronger the evidence. Hypotheses are confirmed individually, however, as the probability of their truth is raised.

The same main insight, that what we are calling "confirmation" and "evidence" vary conceptually and numerically, is perhaps best illustrated by so-called "crucial experiments" in the history of physics. No one now thinks that there are any *crucial* experiments in the sense that a single experiment can demonstrate definitively either the truth or falsity of a hypothesis or theory as against its available rivals. But there are nonetheless "crucial experiments" that allow us to discriminate between

---

[16]Luborsky et al., claim that "there is much evidence already for their mostly good level of efficacy". As noted, many, perhaps most, people talk this way. But it should be clear that the data do not provide evidence that any one of them is particularly efficacious.

[17]Those tempted to draw the conclusion that psychotherapies are "pseudo-sciences" need to be reminded of the fact, given wide publicity by the former head of Sloan-Kettering Cancer Institute, Lewis Thomas, that the body cures most of the diseases by which it is attacked, without any medical intervention. See Thomas (1974).

[18]Julian Reiss, to provide still another example, runs what we call "evidence" together with what we call "confirmation" in the usual way. He writes, "[a] good theory of evidence should be a theory of both support and warrant". As he goes on to explain, "[s]upport pertains to the gathering of facts, warrant to the making up of one's mind. Gathering of facts and making up one's mind are different processes...[But] we cannot have warrant for a hypothesis without having support" (Reiss 2015, pp. 34–35). On his view, one cannot have "justification" without "evidence," the conflation that we are attempting to undo. Along the same lines, he conflates "gathering facts" with "having evidence," when the two activities are to be rather sharply distinguished.

otherwise equally well-confirmed alternatives, providing very strong evidence for one as against the other.

One notable example has to do with the wave and particle theories of light. The particle theory preceded Newton, but he provided it with a great deal of empirical support by way of his prism experiments and of course lent it his immense authority. But for a variety of reasons, the theory that light consists not of a stream of particles but of waves had gained serious adherents by the first years of the 19th century. Both were confirmed in the sense that a number of widely-observed data could be explained satisfactorily on either; given the data, it was more probable than not that each was true. The problem was to devise a "crucial experiment" the result of which was much more likely on one theory than the other. In 1818, the French Academy of Sciences held a competition to elicit such an experiment. The great mathematician-physicist Poisson quickly deduced from Fresnel's version of the wave theory that the shadow of a disk in a beam of light should have a bright spot in its center. Poisson was a partisan of particles, and he confidently predicted that no such spot would be observed. In fact, another French physicist, Arago, carried out the experiment at once and observed the spot, an observation quickly replicated by many others. On the wave theory, the existence of the spot is, as Poisson showed, highly likely, on the particle theory its likelihood approaches zero. Thus its observation provided strong evidence for the wave over the particle theory. Of course, this was not the end of the matter; efforts to support one or the other theory of light, and to somehow combine or replace them, continue.

This is what evidence (or any similarly-named concept) does, allows us to discriminate in a perfectly objective way between hypotheses which may otherwise be equally well-confirmed. As we have seen, there are cases in which there is no evidence favoring one well-confirmed hypothesis over another, "Dodo-bird cases", and cases in which there is strong evidence for one of a pair of more or less equally-confirmed hypotheses, "Poisson cases". As we will go on to illustrate, there are also paradigm cases in which evidence and confirmation vary dramatically, other cases in which both are high and strong, and still others in which both are weak. What the example also shows is that evidence is always conditional on the data *and* the models. On our evidential approach, it should be emphasized, the data may be misleading in the sense that they may be error-prone, evidence is always for one specific model vis-à-vis another specific model, and there is always a better model out there to be explored.

Now down to the work of making the key insight more precise and defending it against alternative accounts of hypothesis-testing.[19]

---

[19]Although the discussion in this monograph is self-contained, readers who would like a more general introduction to some of the notation and basic concepts of elementary probability theory and statistics might look at (Bandyopadhyay and Cherry 2011).

References

Bandyopadhyay, P. and Cherry, S. (2011) Elementary probability & statistics: a primer. In Bandyopadhy and Forster (Eds.), *Handbook of statistics*. Amsterdam: Elsevier,

Bernecker, S., & Dretske, F. (Eds.). (2000). *Knowledge*. New York: Oxford University Press.

Bonjour, L. (1985). *The structure of empirical knowledge*. Cambridge: Harvard University Press.

Bovens, L., & Hartmann, S. (2004). *Bayesian epistemology*. Oxford: Oxford Univeristy Press.

Brody, B. (Ed.). (1970). *Readings in the philosophy of science*. NJ: Prentice-Hall.

Cox, D. (2006). *Principles of statistical inference*. Cambridge: Cambridge University Press.

Crupi, V. 2013/2014. Confirmation. *Stanford Encyclopedia of Philosophy*. http://plato.stanford.edu/entires/confirmation.

DiFate, V. 2007. Evidence. *Internet Encyclopedia of Philosophy*. www.iep.utm.edu/evidence.

Goldman, A. (1967). A causal theory of knowing. *Journal of Philosophy, 64*, 352–372.

Hacking, I. (1975). *The emergence of probability*. Cambridge: Cambridge University Press.

Hempel, C. (1965). *Aspects of scientific explanation*. New York: The Free Press.

Hume, D. 1888/1739-1740. *A Treatise on Human Nature*. Oxford: Clarendon Press.

Kim, J. (1988). What is naturalized epistemology? In J. Tomberlin (Ed.), *Philosophical perspectives, epistemology*. Atascadero, CA: Ridgeview Publishing.

Luborsky, L., Rosenthal, R., Diguer, L., Andrusyna, T., Berman, J., Levitt, J., et al. (2002). The dodo bird verdict is alive and well - mostly. *Clinical Psychology: Science and Practice, 9*(1), 2–12.

Oliver, A. 1999. A Few More Remarks on Logical Form. *Proceedings of the Aristotelian Society* (New Series) XCIV: 247–272.

Reiss, J. (2015). *Causation, evidence, and inference*. New York: Routledge.

Rosenzweig, S. (1936). Some implicit factors in diverse methods of psychotherapy. *American Journal of Orthopsychiatry, 6*, 412–415.

Salmon, (1975). Confirmation and Relevance. In Maxwell and Anderson (Eds.), *Induction, probability, and confirmation*, Minnesota studies in the philosophy of science, (Vol. 6) pp 5–36, Minneapolis: University of Minnesota Press.

Sosa, E., Kim, J., Fantl, J., & McGrath, M. (Eds.). (2010). *Epistemology: an anthology*. New York: Blackwell.

Thomas, L. (1974). *Lives of a cell: notes of a biology watcher*. New York: Penguin Books.

Chapter 2
Bayesian and Evidential Paradigms

Abstract The first step is to distinguish two questions:

1. Given the data, what should we *believe*, and to what degree?
2. What kind of *evidence* do the data provide for a hypothesis H_1 as against an alternative hypothesis H_2, and how much?

We call the first the "confirmation", the second the "evidence" question. Many different answers to each have been given. In order to make the distinction between them as intuitive and precise as possible, we answer the first in a Bayesian way: a hypothesis is confirmed to the extent that the data raise the probability that it is true. We answer the second question in a Likelihoodist way, that is, data constitute evidence for a hypothesis as against any of its rivals to the extent that they are more likely on it than on them. These two simple ideas are very different, but both can be made precise, and each has a great deal of explanatory power. At the same time, they enforce corollary distinctions between "data" and "evidence", and between different ways in which the concept of "probability" is to be interpreted. An Appendix explains how our likelihoodist account of evidence deals with composite hypotheses.

Keywords Confirmation · Evidence · Bayesianism · Likelihoods · Interpretations of probability · Absolute and incremental confirmation · Lottery paradox · Composite hypotheses

© The Author(s) 2016
P.S. Bandyopadhyay et al., *Belief, Evidence, and Uncertainty*,
Philosophy of Science, DOI 10.1007/978-3-319-27772-1_2

Two Basic Questions

Consider two hypotheses, H_1, that a patient suffers from tuberculosis, and H_2, its denial. Assume that a chest X-ray, administered as a routine test for the presence of tuberculosis, comes out positive. Given the datum that the test is positive, and following the statistician Richard Royall's lead, one could ask three questions:[1]

1. Given the datum, what should we *believe*, and to what degree?
2. What kind of *evidence* does the datum provide for H_1 against an alternative hypothesis H_2, and how much?
3. Once questions 1 and 2 have been answered, what should we *do*?

We call the first question the *confirmation question*, the second the *evidence question*, and the third the *decision question*.[2] Like Royall, we think that they are very different questions. Our concern in this monograph is with the first two.[3] A number of answers have been given to each. We want to use two of these answers to make clearer an intuitive distinction between confirmation and evidence which the questions presuppose, and then to show how this distinction both advances our understanding of uncertain inference and provides solutions to notable epistemological puzzles. To this end, and for illustrative purposes, we draw on the Bayesian and Likelihood statistical paradigms, the first to make precise a conception of confirmation, the second of evidence. Each is at least somewhat familiar to many philosophers and scientists, and all statisticians. In Chap. 3, we will show in more

---

[1]See Royall (1997). That the distinction between belief and evidence questions is pre-theoretically intuitive is underlined by the fact that Royall himself is a Likelihoodist who eschews any reference to an agent's subjective degrees of belief (he is, however, a Bayesian in regard to the decision question). Despite the philosophical differences that one or another of us has with him, our monograph owes a great deal to his work. See in particular Royall (2004).

[2]For a Bayesian response to Royall's three questions, see Bandyopadhyay (2007).

[3]Mark Kaplan (1996) is one of the very few philosophers to take note of the confirmation/evidence distinction (p. 25, footnote 32), but his argument for making it seems to involve no more than a reference to Nelson Goodman (quoted on p. 26, footnote 34), to the effect that "[a]ny hypothesis is 'supported' by its own positive instances; but support … is only one factor in confirmation." Kaplan's own very interesting and extensive account of evidence is itself generally "Bayesian" and makes no use of likelihoods. Goodman thinks that since incompatible and arbitrary hypotheses are "supported by the same evidence," there must be another "linguistic" (data-independent) factor involved in "confirmation." As we will see in Chap. 9, Goodman's well-known "grue paradox," which he uses to argue for this claim, depends on running "confirmation" and "evidence" together. Others who have made a confirmation/evidence distinction include Ellery Eells and Branden Fitelson in their (2000) and Malcolm Forster and Elliott Sober in their (2004). Forster and Sober are neither Bayesians nor Likelihoodists, which fact underlines our claim that the distinction is pre-theoretical, that is to say, statistical-paradigm independent.

detail how confirmation and evidence differ, and as a corollary how "data" are to be distinguished from "evidence." In Chap. 4 we consider, and then reject, four very general objections that have been made to the kind of account we set out in this monograph. In Chaps. 5, 6, 7 and 8 we will discuss some modifications of and alternatives to them.

Probabilities and Beliefs

On the Bayesian way of construing what we (and not Royall) call the "confirmation question", the answer has to do with *beliefs* and with *belief probabilities*. On it, *data confirm (disconfirm) a hypothesis just in case they raise (lower) our degree of belief in it*. In John Earman's vocabulary, we "probabilify" hypotheses by the data.[4]

There are at least three reasons for adopting a belief-probabilistic approach, over and above various difficulties with the alternatives to doing so.

First, such an approach reflects the inductive character of the relation between data and hypotheses. As we emphasized in Chap. 1, hypotheses "go beyond" the data enlisted to support them in the sense that all of the data gathered might be correct and yet the hypothesis, on the basis of further such data, be false.[5] One way in which to put this is to say that hypotheses are never more than probable given the data.

Second, the mathematical theory of probabilities is well understood, universally accepted, and precise. It allows for the methodical re-adjustment of the probabilities of hypotheses over time in the light of new or different data, and thus captures the non-monotonic character of inductive inference. In this way it also allows us to rank-order hypotheses with respect to the data that have been gathered or observed.

Third, the use of probabilities affords simple and straightforward measures of *relevance*. Thus data D are relevant to the confirmation of a hypothesis H just in case $\Pr(H \mid D) \neq \Pr(H)$. Similarly, an auxiliary hypothesis H' is useful in predicting data from a target hypothesis H just in case $\Pr(D \mid H \& H') \neq \Pr(D \mid H)$, and so on for many other examples.

There is more controversy in construing probabilities in this particular context as degrees of "belief", something in the heads of agents, and in this sense "subjective." The construal can be bolstered by two considerations.[6]

---

[4]Earman (1992).

[5]Although most of the hypotheses we will use to illustrate our argument do not take the form of universal conditionals, "All A are B," it is especially clear in their case that the claims they express typically outrun the inevitably finite data gathered to support them.

[6]Understanding probabilities as degrees-of-belief and connecting them to confirmation has a long history. See, for example, (Keynes 1921, pp. 11–12). Carnap, too, thought that inductive probability, i.e., the probability of the conclusion of an inductive argument given its various data premises, is "ascribed to a hypothesis with respect to a body of evidence... To say that the hypothesis h has the probability p (say 3/5) with respect to the evidence e, means that for anyone to

First, confirmation has to do with something like a "logical" (although not deductively valid) relation between data and hypotheses, i.e., it has an inferential structure. But inferential structures are propositional in character, they express relations between something like sentences. Beliefs on the usual account are propositional attitudes, they have the sort of sentential form that allows them to enter into inferences.[7] Beliefs like propositions are the bearers of truth and falsity, and probabilities attach to both.

Second, Royall's "what to do?" question presupposes that the first be expressed in terms of degrees of belief. For on the usual Aristotelian account, we act not on the basis of what is true or somehow established, but on the basis of what we believe to be true or established. For Bayesians, confirmation is linked to action by way of the concept of belief.[8]

To say that this account of confirmation is probabilistic, and that probabilities in connection with it are identified with degrees of belief, is to say that this account of confirmation is generally "Bayesian."[9] What makes it more specifically Bayesian is that central importance is accorded to Bayes Theorem, a way of conditioning

---

(Footnote 6 continued)

whom this evidence and no other relevant knowledge is available, it would be reasonable to believe in *h* to the degree *p*, or, more exactly, it would be unreasonable for him to bet on *h* at odds higher than [*p(h)/p(1-h)*].... Thus inductive probability measures the strength of support given to *h* by *e* or the *degree of confirmation* of *h* on the basis of *e* (Carnap 1950, p. 441) As Skyrms (1986, p. 167) summarizes the situation, the concepts of inductive and epistemic (which, as we saw in Chap. 1, applies to statements rather than arguments) probabilities were introduced ... as numerical measures grading degree of rational belief in a statement and degree of support the premises give its conclusion.... Why should epistemic and inductive probabilities obey the mathematical rules laid down for probabilities and conditional probabilities? One reason that can be given is that these mathematical rules are *required* by the role that epistemic probability plays in rational decision" (our italics). James Hawthorne helped prepare this brief history. See his (2011). For our purposes, it is as important to note that neither Keynes, nor Carnap, nor Skyrms distinguishes between confirmation and evidence (as the title of the selection from Carnap's work in Achinstein 1983, "The Concept of Confirming Evidence," makes clear).

[7]Although it is not needed for our argument, it is worth mentioning that a confirmation-generalization of logical entailment has been worked out by Crupi et al.(2013).

[8]On the traditional account of voluntary action, an action is voluntary just in case it is performed by a rational agent on the basis of her desires and beliefs. For Skyrms and other Bayesians, "rational agency" requires at a minimum that the agent's beliefs conform to the rules of the theory of probability.

[9]There are other ways in which to model belief within the context of a confirmation theory, for example, the Dempster-Shafer belief function. See Shafer (1976). Since the probability-based account is well-known and has a long tradition, we are resorting to it.

degrees of belief on the data gathered. At any time t_i, Bayes Theorem (to be described very shortly) tells us to what extent it is reasonable to believe a particular hypothesis given the data. On this account of confirmation, an agent should change her degree of belief in a hypothesis H from t_i to t_{i+1} by the amount equal to the difference between the posterior probability of H, $\Pr(H \mid D)$, and $\Pr(H)$, its prior probability.

Bayesian confirmation theory is "subjective" in the sense that all probabilities are identified with degrees of belief and confirmation with the way in which new data raise the probability of initially-held or prior beliefs. That said, many philosophers believe that some Bayesian approaches are more "objective" than others, depending on the constraints put on the determination of prior probabilities.[10] Those who reject a Bayesian approach in toto do so entirely on the basis of such "subjectivity", large or small, which in their view is incompatible with the objectivity of science, although they also allege that subjective belief leads to paradox. We will try to unravel the alleged paradoxes later in this chapter and in Chap. 9. But we will not try to defend the Bayesian approach against every challenge. We think that the rule of conditional probability on which it rests makes precise the notion of learning from experience, and that learning from experience is the intuitive basis of all confirmation. Our use of the Bayesian approach in this monograph, however, is purely instrumental, and is intended to make the distinction between confirmation and evidence as sharp as possible.

Bayesian Confirmation

The account of confirmation we take as paradigm involves a relation, $C\,(D,\,H,\,B)$ among data D, hypothesis $H,$ and the agent's background information $B.$[11] However it is further specified, it is modeled on the basic rules of probability theory including the rule of conditional probability, together with some reasonable constraints on one's a priori degree of belief in whatever empirical hypothesis is under consideration. If learning from experience is to be possible, one of these constraints is that the agent should not have an a priori belief that an empirical hypothesis is true to degree 1, i.e., with full certainty, or 0, in which case it would have to be self-contradictory. This said, the agent learns from experience by up-dating her

---

[10]See Bandyopadhyay and Brittan (2010).

[11]Except when it is important, we leave out reference to the background information in what follows.

degrees of belief that hypotheses are true by conditionalizing on the data as she gathers them, i.e., in accord with the principle, derivable from probability theory, that $\Pr(H \mid D) = \Pr(H)\Pr(H \& D)/\Pr(D)$. Assuming that $\Pr(D) \neq 0$, her degree of belief in H after the data are known is given by $\Pr(H \mid D)$. Thus, D confirm H if and only if $\Pr(H \mid D) > \Pr(H)$. Call this the *Confirmation Condition*.[12] It is qualitative, i.e., compares the probabilities of a hypothesis before and after data have been collected, the intuitional basis of this conception of confirmation. This definition rests, as do most Bayesian conceptions of confirmation as a probability measure, on the following principle: for any H, D_1, D_2,, the confirmation (disconfirmation) of H in the light of D_1 is greater (less) than the confirmation(disconfirmation) of H in the light of D_2 just in case $\Pr(H \mid D_1) > (<) \Pr(H \mid D_2)$.[13] This principle makes explicit that as the probability of the hypothesis increases (decreases) as a result of further data-gathering, so too does its degree of confirmation (disconfirmation). A quantitative notion of confirmation of a hypothesis at any given time, is measured, for instance, in terms of the difference between its prior and posterior probabilities.[14] A hypothesis is always confirmed to some degree if the confirmation condition is satisfied. Whether it is "low" or "high" depends on the particular confirmation measure chosen,[15] the implicit standards of particular scientific communities, and the purposes of the investigator. On its Bayesian reading, the posterior probability of a hypothesis H equals its prior probability multiplied by the probability of D given H, $\Pr(D \mid H)$, divided by the marginal probability of D, $\Pr(D)$:

$$\Pr(H|D) = \Pr(H)\Pr(D|\mathrm{H})/\Pr(D) \tag{1}$$

---

[12]Or as it is sometimes called, "the positive relevance condition." See Salmon (1983) for an extended argument in behalf of the primacy of this condition in an analysis of confirmation.

[13]Following Crupi et al. (2013). The article includes a long list of Bayesians who subscribe to this principle.

[14]Clark Glymour (in an e-mail comment to us) and Peter Achinstein (2001, especially Chap. 4) object that this sort of account has a counter-intuitive consequence, that the same data could confirm incompatible hypotheses to different degrees. But so long as our assignments of degrees of belief are consistent, i.e., do not violate the rules of probability theory, it is possible to be rationally justified in believing incompatible hypotheses to different degrees on the basis of the same data.

[15]That is to say, we use this as an exemplary measure of degree of confirmation. Many others are possible. See Fitelson (1999), for a discussion of the sensitivity of confirmational values to the measure used. We believe that the choice of a specific confirmation measure depends on the type of question one is asking. The same idea, that the type of question asked determines the measure chosen, applies to the evidence question as well. Although we have adopted the likelihood ratio to weigh evidence, different evidential measures would be required *if* we were to ask a different set of evidential questions. See Taper et al. (2008) and Chap. 5 for further discussion of alternative measures.

$\Pr(H \mid D)$ is also called the conditional probability of H given D. The prior probability of a hypothesis represents the agent's degree of belief that the hypothesis is true before (i.e., prior to) new data bearing on the hypothesis have been gathered. This agent-relative prior probability component of the definition is the most controversial element in the application of Bayes' Theorem, and will be discussed in more detail later.

The quantity $\Pr(D \mid H)$ is often referred to in the *philosophical* literature as the likelihood. While numerically the likelihood of the hypothesis given the data is equal to the probability of the data given the hypothesis, likelihood and probability are not the same thing; likelihood is considered a function of the hypothesis, whereas the probability is considered a function of the data. We here adopt the common philosophical notation of denoting the likelihood by $\Pr(D \mid H)$ rather than the common statistical notation of L $(H; D)$,[16] but do not mean to imply that the hypothesis H *needs* to be considered a random variable.

The likelihood function provides a tool, through the likelihood ratio, to answer the question, "How much support for a hypothesis is there in the data relative to another hypothesis?" The likelihood function is an important tool for Bayesians and non-Bayesians alike, but too rarely accorded the kind of importance that we do here.

The final element of Eq. 1 is $\Pr(D)$. This is calculated as the marginal probability of the data over the alternative hypotheses, that is, the probability that D would obtain, averaged over H and $\sim$H:[17]

$$\Pr(D) = \Pr(H)\Pr(D|H) + \Pr(\sim H)\Pr(D|\sim H). \tag{2}$$

---

[16]In the eyes of many statisticians this notation signals the difference between "probability" and "likelihood," as two different concepts. More than a simple notational difference is involved. The " $\mid$ " notation indicates conditioning on a random variable, i.e., in the case of $\Pr(D \mid H)$ the hypothesis is a random variable, while $\Pr(D; H)$ indicates that the data are conditioned on a variable that is considered fixed. The first is fundamentally Bayesian, the second is fundamentally evidentialist.

[17]For convenience, hypotheses are most often presented as exhaustive pairs, H and H, but in theory the list of hypotheses considered is not limited to such pairs. It is difficult, among other reasons, to compute the probability of the data under the catch-all hypothesis [$\Pr(D \mid \sim$H), and in consequence it is difficult to calculate the posterior probability of the catch-all. This difficulty then extends to comparing the posterior probability of the catch-all with the posterior probabilities of other hypotheses. We avoid such difficulties by confining our discussion to *simple* hypotheses. It might be added here that on the present account of what is often called "incremental confirmation," the Special Consequence Condition does not hold see Salmon (1983, p. 106). To handle objections in this connection, Kotzen has produced a principle which he calls "Confirmation of Dragged Consequences:" If [$\Pr(H \mid D) > \Pr(H)$, and H_1 entails H_2, and $\Pr(H_2) < \Pr(H_1 \mid D)$] then $\Pr(H_2 \mid D) > \Pr(H_2)$. See Kotzen (2012).

Evidence and Likelihood Ratios

As we understand it, evidence invariably involves a comparison of the merits of two hypotheses,[18] H_1 and H_2 (possibly, but not necessarily, $\sim H_1$) relative to the data D, background information B, and auxiliaries A.[19] This is to say that we distinguish "evidence" from "data." All evidence depends on data. But data constitute evidence only when, in context, they serve to distinguish and compare hypotheses. Four preliminary points might be made in this connection.

First, it is a commonplace that not all data constitute evidence. Whether they do so or not depends on the hypothesis being tested, whether the data "fit" or are relevant to appraising the hypothesis. Whether data constitute evidence is a matter of context. Data themselves, which are often taken as a first approximation as the reports of observations and the results of experiments, are in this sense context-free.

Second, data are paradigmatically invoked as evidence in the history of science in a comparative context; Galileo's sighting of Jupiter's moons was rightly taken as evidence for the truth of the Copernican as against the Ptolemaic hypothesis.

Third, a comparative account of evidence meets the demand of one key goal of science. The goal is to understand and explain natural phenomena. To do so, scientists propose descriptive-explanatory models of these phenomena, and seek better and better models as closer and closer approximations to the truth. We think that there is in some meaning of the words, truth or reality which we study with the help of models. But, none of the models are true since they all contain idealizations which are false. Our evidential framework is designed to quantify in what sense and by how much one model is a superior approximation to the truth about natural phenomena than another.

Fourth, a distinction between data and evidence first allows us to understand and then helps resolve many if not all of the difficulties that beset traditional theories of confirmation. In Chap. 9, for example, we take up "the old evidence problem" for subjective Bayesian accounts of confirmation. Unless a distinction between data and evidence is made, the problem is trivial. To the question "does old evidence provide evidence for a new theory?" the obvious answer is "yes, of course." But the

---

[18]Statisticians prefer to use the term "models," by which they mean statistical models that allow for quantitative testing vis-à-vis the data, rather than "hypotheses" in this connection. Although we mostly stick to the more general (albeit vaguer) philosophical usage, the difference in meaning between the terms is important. As we emphasize at the end of this chapter in connection with various interpretations of probability, a hypothesis is a verbal statement about a natural state or process, a model is a mathematical abstraction that captures some of the potential of the hypothesis. Although we occasionally use the terms interchangeably, it matters a great deal whether one has models or hypotheses in mind when it comes to the correct characterization of the hypothesis-data and model-data relationships.

[19]Again in what follows, and except where it matters, we will not include reference to A or B in our formulations.

question is no longer trivial when it is re-phrased: "do old data provide evidence for a new theory?" The answer to this second question requires serious reflection.[20]

On our account of evidence, a model or hypothesis is a defined data-generating mechanism. Observed data support model$_1$ over model$_2$ if the data that would be generated by model$_1$'s mechanism are by some measure more similar to the observed data than the data that would be generated by model$_2$. Thus on our account, data provide evidence for one hypothesis against its alternative *independent of what* the agent *knows or believes* about either the available data or the hypotheses being tested.[21] A subtle but important distinction in this connection is between knowing that the data are available and knowing the probability of the data. One can know how probable observations of the data are without knowing whether the data have actually been observed. For example, one could calculate the probability of obtaining 100 tails out of 100 flips of a coin on the hypothesis that the coin is biased towards tails with a probability of 0.9. This differs from asserting that we know on the basis of the data that out of 100 flips a coin has landed tail-side up 90 times.

One final preliminary. We have assumed for the sake of clarity and convenience that the hypotheses in our schematic examples are simple and not complex. Error-statisticians object that clarity and convenience have little to do with it; we are forced to make this assumption in the case of our account of evidence because it cannot deal with composite hypotheses.[22] The issues involved are technical, and for that reason we have put our discussion of them in an Appendix to this chapter. Suffice it to say here that this objection can be met.

The Evidential Condition

Now back to our characterization of evidence. It is made precise in the following equation:[23]

D is evidence for $H_1\&B$ as against $H_2\&B$ if and only if $\mathrm{LR}_{1,2}$
$$= \left[\frac{\Pr(D|H1\&B)}{\Pr(D|H2\&B)}\right] > 1$$

---

[20]Robert Boik pointed this out to us.

[21]It thus stands in sharp contrast to the well-known position of Williamson (2000), whose summarizing slogan is "a subject's evidence consists of all and only the propositions that the subject knows." Williamson is not analyzing the concept of "evidence," or more precisely "evidence for a hypothesis or theory," but the concept of "evidence for a subject." This concept is important in classical epistemology, but has little use, or so we believe, as applied to scientific methodology or theory assessment. For us, evidence is evidence, whether or not the subject knows it, and conversely, whether or not the subject knows something does not thereby qualify it as "evidence" for any particular hypothesis.

[22]See Mayo (2014).

[23]We use "LR" rather than "BF" in what follows to underline the fact that our account of evidence is not in any narrow sense "Bayesian.".

Call this the *Evidential* condition. It serves to characterize data which, in context, play an evidential role. But a quantitative measure is immediately possible by taking account of the numerical ratio of the two hypotheses. Note in this connection that if $1 < LR \leq 8$, then D is often said to provide "weak" evidence for H_1 against H_2, while when $LR > 8$, D provides "strong" evidence for H_1 over H_2 (Royall 1997). This cut-off point is sometimes determined contextually by the relevant scientific communities and may vary depending on the nature of the problem confronting the investigator, but it follows a statistical practice common among investigators.[24] It follows from the Evidential Condition that the range of values for the LR can vary from 0 to ∞ inclusive.[25]

Since Bayesians by definition do not assign the end-point probabilities 0 or 1 to any empirical proposition, it follows that an agent's (subjective) degree of belief in the hypotheses over which she has distributed prior probabilities does not affect whether D is evidence for H_1 as against H_2.[26] This is especially evident if we assign the extreme probability 0 to H_1. In that case, the LR for H_1 against H_2 relative to D becomes $Pr(D \mid H_1)/Pr(D \mid H_2) = [Pr(D \ \& \ H_1)/Pr(H_1)]/ [Pr((D \ \& \ H2)/Pr(H_2)] = 0/0$, which is undefined.[27]

We have argued that evidence has to do with the degree to which data help distinguish the merits of competing hypotheses. Several measures are available to capture the resulting evidential strength of one hypothesis as against another. We have fixed on ratios of likelihood functions as our exemplar in this monograph for three reasons. First, it is the most *efficient* evidential function in the sense that we can gather strong evidence for the best model with the smallest amount of data.[28] Second, the LR brings out the essentially comparative feature of evidence in a clear and straightforward way. Third, and as we have pointed out, it is a measure of

---

[24]Royall (1997) points out that the benchmark value = 8, or any value in that neighborhood, is widely shared. In fact, the value 8 is closely related to the Type I error-rate 0.05 in classical statistics and to an information criterion value of 2. See Taper (2004) and Taper and Ponciano (2016) for more on this issue.

[25]One might argue that the posterior-prior ratio measure (PPR) is equal to the LR measure and therefore precludes the necessity of a separate account of evidence. But the objection is misguided. The LR is equal to $Pr(H \mid D)/Pr(H)$ *only when* $Pr(D)/Pr(D \mid \sim H)$ is close to 1. That is, $[Pr(D \mid H)/Pr(D \mid \dashv H)] = Pr(H \mid D)/Pr(H) \times [Pr(D) \times Pr(D/\dashv H)] \approx 1$. Otherwise, the two measures, LR and PPR, yield different values, over and above the fact that they measure very different things.

[26]Berger (1985, p. 146). Levi (1967) also emphasizes the objective character of the likelihood function.

[27]We are indebted to Robert Boik for this clarification.

[28]See Lele (2004) for the proof. It is worth noting the comments of a referee on a similar claim in another paper of ours: "...if in the end we hope for an account of evidence in which evidence gives us reason to *believe*, it is totally unclear why efficiency in the sense described would be taken as a sign of a meritorious measure of evidence." One of the principal aims of this monograph is to disabuse its readers of what we call the "true-model" assumption, that evidence gives us reasons to believe that a hypothesis is true. In our view, evidence provides information about a hypothesis on the basis of which we can make reliable inferences and not reasons to believe that it is true. Since evidence is a way of identifying and assessing such information, efficiency in the sense indicated is indeed a meritorious property of it.

evidence embraced by such otherwise diverse approaches to statistical inference as Bayesian and Likelihoodist (Evidentialist) insofar as we confine ourselves to simple statistical hypotheses. We are in good company.

We wish to be clear, however, that the likelihood ratio is a special case of a rich family of evidence measures or functions.[29] In principle, an evidence function, Ev (D, M_1, M_2), is the difference between the statistical distance[30] between the probability distributions generated by each of the models, M_1 and M_2, and an estimate of the "true" underlying distribution based on D. That is to say,

$$Ev(D, M_1, M_2) = n(SD(\hat{\tau}_D, M_1) - SD(\hat{\tau}_D, M_2)).$$

where n is the sample size, SD(•,•) is a statistical divergence (distance) between two probability distributions indicated by place-holders, $\hat{\tau}_D$ is the estimate of the "true" distribution based on the data D. For discrete distributions, this could be simply the proportion of observations in each discrete data category. For continuous distributions, it will be some smoothed estimate such as a kernel density estimate (Silverman 1986). The inclusion of n in the formula for evidence conforms to the intuitive expectation that the strength of evidence should increase with increased amounts of data.[31]

The most famous statistical distance is the Kullback-Leibler Distance (KLD), the expected log likelihood ratio between two distributions. For discrete distributions, P and Q with categories indexed by I, this is given as: $KLD(P, Q) = \sum_i P_i \cdot [\log(P_i) - \log(Q_i)]$

It is easy to show that the log (LR) is the KLD-based evidence function for comparing simple hypotheses. Similarly, the differences of information criterion values are KLD-based evidence functions for cases where the models compared differ in the number of parameters estimated.[32]

---

[29]What follows is drawn from Lele (2004). We go into the details, technical as some of them are, simply because evidence functions are so much less familiar than confirmation functions.

[30]In fact, evidence functions only use the weaker criterion of divergences (statistical distances are divergences with some additional constraints). A divergence quantifies the dissimilarity between two probability distributions. The statistical literature is quite loose in its use of the terms "divergence," "disparity," "discrepancy," and "distance." We use "distance" rather than the more general term "divergence" because "distance" is more intuitively evocative. Good discussions of statistical distances can be found in Lindsay (2004) and Basu et al. (2011).

[31]Although it does not do so in a linear fashion.

[32]The technical definition of evidence functions (Lele 2004) includes a consistency requirement (i.e., the probability of correctly selecting the best approximating model must go to 1 as sample size goes to infinity). Thus only "order consistent" information criteria such as Schwarz's criterion (variously denoted as the SIC or BIC) can be used to construct evidence functions.

A large number of potential statistical distances could be used to construct evidence functions in response to different types of evidence question. For example, one alternative to the KLD is the Hellinger distance.[33] Evidence functions based on different distances will have different statistical properties. For instance, while KLD-based evidence functions will be maximally efficient, Hellinger-based evidence functions will be more resistant to the presence of outliers. But none of the following discussion turns on these technical details.

Absolute and Incremental Confirmation

An important objection that has been made to our account is that its main theme, that confirmation and evidence should be distinguished, has already been developed in the literature in terms of confirmation alone.[34] The objection is that both (Carnap 1950) and (Salmon 1983) long ago made a parallel distinction between "absolute" and "incremental" confirmation which does all of the work that ours does without recourse to any distinct notion of evidence. On the absolute concept, sometimes also called "the high probability requirement," the data confirm H if and only if the probability of H given D exceeds some suitably high threshold, say, 0.9 (or minimally >0.5). It thus picks up on an ambiguity in the word "confirmed" already introduced. In certain contexts, the word connotes something like "well confirmed" or "put to rest", as in "Well, we certainly managed to confirm that guess." In this respect it is to be contrasted with the incremental way in which we confirm hypotheses or for that matter hunches, gathering evidence as we go along, becoming more and more sure that our hunch was right or wrong.

On the incremental conception of confirmation favored by most Bayesians, the data D confirm H if and only if the data raise the probability of H relative to its prior probability. It is how we understand confirmation here. Although there are important differences between the incremental and absolute conceptions, neither can be used to explicate, still less is tantamount to, anything like the notion of evidence that we and many others draw on. First, and as we will see in more detail in the next chapter, strong evidence does not entail a high degree of confirmation, either absolute or incremental. Second, confirmation of both varieties is sensitive to an agent's prior probability distribution and endorses a rule for up-dating degrees of belief, whereas evidence is insensitive to prior probabilities, characterizes the relation between data and hypothesis regardless of whether the agent knows or believes that either data or hypothesis is probable, let alone certain or true, and indicates no way in which to up-date degrees of belief. Third, and unlike our

---

[33](HD): $HD(P,Q) = {1}/{\sqrt{2}}\sqrt{\sum_i \left(\sqrt{P_i} - \sqrt{Q_i}\right)^2}$ for discrete distributions.

[34]Two previous readers raised this objection. We are assuming the standard account of absolute confirmation. Please see the following footnote for more on this point.

characterization of evidence, absolute confirmation is, as its name suggests, absolute; if one restricts what counts as evidence to data that confirm a hypothesis absolutely, there is no way in which to determine whether some evidence is stronger than others, still less a way to quantify its strength.[35] Given a prior $\Pr(H) = 0.2$ and a posterior $\Pr(H \mid D) = 0.901$, H is intuitively much more strongly confirmed than in a parallel situation in which $\Pr(H) = 0.899$ and $\Pr(H \mid D) = 0.90$, but the notion of absolute confirmation is unable to capture this intuition. Still worse, it undermines it.

Finally, and perhaps most importantly, the concept of absolute confirmation, unlike the more standard conception of evidence, applies to hypotheses considered singly and not pair-wise, i.e., it does not necessitate a separation and comparison between rival hypotheses which is, we contend, at the center of what we should expect a measure of evidence to provide; again as in courts of law, evidence is what distinguishes the guilty from the blameless and indicates which data determine who is guilty and who not, and to what degree.

Quite apart from its failure to capture the intuitive concept of evidence, the notion of absolute confirmation has its own difficulties. We mention two because each throws more light both on the incremental conception and on the distinction with evidence that we use it to make.

One difficulty with absolute confirmation is that it runs directly into the lottery paradox.[36] Suppose a fair lottery with a thousand tickets. Exactly one ticket will win and, since the lottery is fair, each stands an equal chance of doing so. Consider the hypothesis, "ticket #1 will not win." This hypothesis has a probability of 0.999. Therefore we have a conclusive reason, on the absolute conception, to believe that it is true. But the same line of reasoning applies to all of the other tickets. In which case, we should never believe the hypothesis that any one of them will win. But we know, given our initial supposition, one of them will win. The paradoxical result can be avoided by denying that any hypothesis is ever absolutely confirmed.[37]

The lottery paradox also exposes one way in which our distinction between (incremental) confirmation and evidence is of real use. Sober uses the lottery paradox to argue for a wholesale rejection of the notion of "acceptance." First, to accept a hypothesis is to have a good reason for believing that it is true (or empirically adequate, or whatever). But the converse does not hold. However good our reason for believing it to be true, however well-confirmed, we might still not accept the hypothesis. We might not accept it because the otherwise confirming

---

[35]There is a non-standard account of "absolute confirmation" in the literature on which it does admit of degrees; on this account a hypothesis is "absolutely confirmed" if it is "confirmed strongly," where "confirmed strongly" can have different degrees. See Eells (2006, p. 144). Our argument depends on the standard account, on which D confirm H "absolutely" just in case $\Pr(H \mid D) = r$, where r is identified as a particular number (or, occasionally, any number greater than 0.5).

[36]First propounded by Kyburg (1961). We follow Elliott Sober's informal statement of it (1993).

[37]Kyburg himself avoids the paradox by denying the uncritical "conjunction principle" on which it rests, that if each of a set of hypotheses is accepted, then their conjunction must be as well.

data are not evidentially significant.[38] They are not evidentially significant, on our characterization, when they fail to distinguish between competing hypotheses. In the lottery case, the likelihoods of all of the competing hypotheses, that is, the likelihood of cashing a winning ticket on the hypothesis that it is not a winning ticket, arc the same. In which case, the data fail to distinguish between the competing hypotheses, in which case they are, in context, evidence for none of them. If acceptance requires evidential significance, as we will in more detail in Chap. 6, then we should not accept any of the hypotheses.

The other difficulty with the absolute conception of confirmation is that it entails what is sometimes called the inconsistency condition: the data can never confirm incompatible hypotheses. But when more than two hypotheses are at stake, the data can and do incrementally confirm more than one, however incompatible they might be.[39]

Our Two Accounts and Interpretations of Probability[40]

Before proceeding further, we need to make explicit what has been implicit from the outset, that adequate accounts of confirmation and evidence presuppose different interpretations of the concept of "probability", and therefore different readings of the probability operator in our various conditions and equations as it is applied to events, hypotheses, and propositions. What is important to our fundamental distinction is not so much the details of these readings, still less the widespread and often intense controversies to which they have given rise, but that some

---

[38]This is our view, not Sober's.

[39]A well-known example was provided by Popper (1959, p. 390). "Consider the next throw with a homogeneous die. Let x be the statement 'six will turn up'; let y be its negation…; and let z be the information 'an even number will turn up'. We have the following absolute probabilities: $p(x) = 1/6$; $p(y) = 5/6$; $p(z) = 1/2$. Moreover, we have the following relative probabilities: $p(x, z) = 1/3$; $p(y, z) = 2/3$. We see that x is supported by z, for z raises the probability of x from 1/6 to $2/6 = 1/3$. We also see that y is undermined by z, for z lowers the probability of y by the same amount from 5/6 to $4/6 = 2/3$. Nevertheless, we have $p(x,z) < p(y, z)$." Popper mistakenly drew the conclusion that there was a logical inconsistency in Carnap's confirmation theory. But the inconsistency follows only if we were to take confirmation in its "absolute" sense, i.e., just in case the data raise the probability of a hypothesis beyond a high threshold. There is no inconsistency if confirmation is taken, as we do, in its incremental sense. See also Salmon (1983, pp. 102−03).

[40]John G. Bennett has drawn our attention to the need to address the varying interpretations of probability involved in our accounts. In a very helpful e-mail communication, Marshall Abrams has underlined some of the difficulties in calculating probabilities from probabilities defined by two different interpretations and in taking as paradigm some of the ways in which the likelihood relationship between data and hypothesis has been construed. As for the first, we offer reasons in what follows for taking posterior probabilities as subjective despite their embedding objective components. As for the second, our account of objective probabilities uses empirical frequencies to estimate and not to "define" probabilities.

of them are generally "subjective" and others "objective", and that there are "objective" ingredients in our account of confirmation as well as in our account of evidence. In our view, one needs to mix and match probabilities of different kinds and not rely on one to the exclusion of the others. As we will now show, to do otherwise would be to seek an overly monistic methodology.

Consider first our account of confirmation. It is informed by a subjective Bayesianism. D confirm H just in case an agent's prior degree of belief in H is raised. The degree to which D confirm H is measured by the extent to which the degree of belief has been raised. The probabilities involved have to do with belief, and are in this sense subjective. But determining an agent's degree of belief in a hypothesis, $\Pr(H \mid D)$ requires determining the probability of D on H, $\Pr(D \mid H)$, which is independent of an agent's belief and is therefore objective.

One possible response to this claim is that the probability function is mistakenly construed as objective; it needs to be construed as a conditional belief-probability in which the probability operator is understood in *subjective* terms. But even otherwise staunch Bayesians are not in full agreement to do so, witness L.J. Savage's admission in his famous 1963 paper with W. Edwards and H. Lindman[41] that likelihoods are "public," implying that they are objective. As Isaac Levi points out in this connection,[42] likelihoods are agent-invariant and fixed, and in this sense objective, and unlike variable conditional probabilities which are agent-dependent.[43] At the same time, subjectivity spreads through a compound probability like falsehood through conjunctions and truth through disjunctions; since the posterior probability, $\Pr(H \mid D)$, is calculated in part on an agent's prior degree of belief it is subjective, as should be expected on a Bayesian account of confirmation.

There is a possible justification from the subjective Bayesian standpoint regarding how one could mix and match those two types of probabilities together. David Lewis argues that when the chance probability of a proposition, A, is available, one needs to set an agent's subjective probabilities, $\Pr(A|M) = \Pr(M)$ where M is a statistical model (Lewis 1980.)[44] Lewis thinks that this alignment of subjective probability with objective chance is possible because of the *Principal Principle*. The idea behind this principle is that when we have the likelihood of a

---

[41]Edwards et al. (1963).

[42]Levi (1967).

[43]This is perhaps clearest when D are entailed by H, for in this case $\Pr(D \mid H) = 1$, regardless of what anyone happens to believe.

[44]We are indebted to both John. G. Bennett and Colin Howson for some clarifications about the need to introduce the *Principal Principle*. For more on the justification of the Principal Principle, see the forthcoming book by Robert Pettigrew, *Accuracy and the Laws of Credence*, Part II of which is devoted to it. We are indebted to Jason Konek for calling our attention to this important discussion.

model given the data (what we have just called "propositions") available we should treat an agent's subjective probability which is a conditional probability of the data given a model to be equal to its objective likelihood. This alignment of the subjective probability with the objective chance helps treat both probabilities, the likelihood functions, and prior probabilities, in the application of the Bayes theorem as subjective.[45]

Consider, second, our evidentialist account. It makes no use of belief-probabilities and in this sense is thoroughly objective. But in a way generally similar to the account of confirmation, it too mixes and matches interpretations of the probability operator in a way that without explication might invite misunderstanding but that, though nuanced, is necessary to get our inference-engine running. The likelihood of a model given the data is an essential component of the account. For simplicity's sake, we identify it with a conditional probability, $\Pr(D_0|M)$, where "D_0" stands for a yet to be realized model-generated datum and M is the generating model. What data will be realized a priori depends on what sorts of models we have proposed. At the initial stage, when no data have yet been realized, the relationship between a model[46] and its unrealized data is deductive. One could come up with an open-ended number of distinct models, M_1, M_2, M_3,...M_k. Let's assume M_1 says that the coin has 0.9 probability to land with heads. Assume further that M_2 says that it has 0.3 probability, M_3 says that it has probability 0.5, and so on. Given these models, before any real data have been observed, each will tell us how probable any set of observations would be under the model. So, the relationship between M_1 and D_0 is completely deductive. So, too, are the relationships between other competing models and $_0$. Now assume that a real coin has been flipped. Further assume that out of 100 flips, 60 of them have landed with heads and the rest are tails. Let's call the data D_1 60 heads and 40 tails. The relationships expressed via $\Pr(D_1|M_1)$, $\Pr(D_1|H_2)$ .....$\Pr(D_1|H_k)$ continue to be deductive. We find that each model tells us how probable those data are under that model, although the probability values will vary from one model to the next.

It is natural to assume that the "propensity" of a model to generate a particular sort or set of data represents a causal tendency on the part of natural objects being modeled to have particular properties or behavioral patterns and this tendency or

---

[45]Given the limited scope of this Monograph, we are not going to evaluate rigorously whether this consideration is well-grounded.

[46]As we noted earlier and for the reason given, we use "model" (less often) and "hypothesis" (more often) interchangeably. But there are crucial differences between them that in the present context might give rise to misunderstandings. A hypothesis is a verbal statement about a state of nature or some natural process. A model is a mathematical abstraction that captures some dimensions of the hypothesis. When we say that the likelihood relationship between data and hypothesis is "logical," this is not in the precise sense in which Keynes or Carnap use the term; it has to do rather with the evaluation of a single hypothesis based on the statistical data. On the other hand, the calculation of the probabilities of the data given two or more models is in our sense of the term "deductive." The central point is that the relationships between data and hypothesis and data and models must be kept distinct; our use of the words "logical" and "deductive" is intended to do so, whatever other connotations those words might carry.

"causal power" is both represented and explained by a corresponding hypothesis. According to the propensity interpretation of probability,[47] the probability of an event is a dispositional property of it.[48] In our case, the probability of a coin's landing heads, if it is flipped, is estimated at 0.6. The theme behind this interpretation is that the probabilistic statement is true because the object in the nature of the case possesses a special sort of dispositional property, called a *propensity*. If a sugar cube would dissolve when immersed, the sugar cube has the dispositional property of solubility. In the same vein, if a coin has a 0.6 probability of landing heads when flipped, the coin is to have a propensity of a certain strength to land heads when flipped. However, we have no way to know for sure what this tendency is except to estimate it through finite empirical frequencies.

The empirical interpretations of probability championed by von Mises[49] and Reichenbach[50] define probability in terms of the limits of relative frequencies; but the empirical sequences we encounter in the world are always finite, and we have often good reason to suppose that they cannot be infinite. Although there are significant mathematical considerations for that kind of idealization in their frequency interpretation, that is, that sequences are infinite, insofar as scientific experiments are concerned this kind of idealization is not implausible. We cannot assume that in a physical experiment we have an infinite number of helium nuclei in various stages of excitation. Nor should we assume in any case that the probability operator is to be "defined" in frequency terms.[51] Rather, finite frequency is no more than a way of estimating propensities in particular cases. Troublesome cases do not undermine an interpretation of "probability" so much as they raise difficulties for approximating values under certain conditions.

In other words, and in our view, it is misleading to say that there are three distinct interpretations of objective probability—deductive, propensity, and finite frequency. Rather, likelihoods allow us to choose models whose deductive probabilities best match natural propensities as approximated by finite frequencies. It is in this way, we think, that they are best to be understood, a case for their agent-independence and objectivity to be made, and a distinction with subjective belief-probabilities to be drawn. As we indicated at the outset, confirmation is in a

---

[47]See Popper (1957). See also Berkovitz (2015) for a defense of the propensity interpretation of probability against the traditional twin criticisms that the explication of propensities is circular and therefore non-informative and that it is metaphysical and therefore non-scientific.

[48]A doctor in Molière's *Le Malade Imaginaire* attributes a "dormitive power" to opium, becoming the object of ridicule then and for the next 300 years. But to attribute a dispositional property is not to provide a causal explanation but to aggregate natural processes which must then be explained at a deeper, in this case pharmaceutical, level of analysis. Sugar has a propensity to dissolve in water, but it has almost no propensity to dissolve in gasoline. This difference can be understood through the chemical structure of sugar, water, and gasoline.

[49]See von Mises (1957).

[50]See Reichenbach (1949).

[51]See Hàjek (1997).

fundamental sense "in the head", evidence in that same sense "in the world." The fundamentally two-fold interpretation of "probability" makes this sense clearer.

The Limits of Philosophical Taxonomy

Philosophical taxonomies play a useful role in locating positions on the conceptual map, drawing attention to their general features, and instructing students. But they can mislead. Although there is a functional, but not conceptual, connection between our characterizations of evidence and confirmation, they are, as just indicated, otherwise quite distinct, and it would be a mistake to bring them under some such common rubric as "Partial-Subjective Bayesianism" or "Probabilistic Evidentialism." Confirmation is to some degree "subjective", involving personal probabilities and the up-dating of beliefs. Evidence is not. Following Kant's guidance, our task here in any case is not to suggest a new statistical or hypothesis-testing paradigm, but to use two already well-established paradigms to draw a distinction which, once drawn and understood, will help chart a new direction in our thinking about uncertain epistemic inference, mitigate, to some extent, the overly adversarial nature of its discussion, and dissolve some traditional epistemological puzzles.

Appendix

A Note on the Likelihoodist Treatment of Simple and Composite Hypotheses

Bayesians use the Bayes Factor (BF) to compare[52] hypotheses (Kass and Rafferty 1995), while others use the likelihood ration (LR) to measure evidence. For simple hypotheses, as in the tuberculosis example discussed near the outset of the next chapter, the Bayes Factor and the Likelihood Ratio are identical; both capture the essential core of our analysis of the concept of evidence. Throughout the monograph we assume that the hypotheses being tested are simple statistical hypotheses, which specifies a single value to a parameter, in contrast to a compound or

---

[52]Glymour (1980), p. 102, rightly draws attention to what he calls "misplaced rigor." But rigor is otherwise indispensable. Error-statisticians have focused their criticisms of the evidential position on an alleged failure to deal with composite hypotheses; see Mayo (2014). This is our reply (following Taper and Lele 2011, Sects. 6 and 7). It is rather technical in character, and does not affect the main line of our argument. We have similarly added technical Appendices to Chap. 6 to reply to another error-statistical criticism of our evidential account, that it needlessly employs multiple models in its account of hypothesis testing, and to Chap. 11, to illustrate some of our main themes in more mathematical detail.

composite hypothesis which restricts a parameter θ only to a range of values.[53] Since some philosophers have claimed recently that the likelihood account of evidence cannot deal with composite hypotheses, it is worth our while to argue why they are mistaken.

Here is a test case:[54]

> [M]edical researchers are interested in the success probability, θ, associated with a new treatment. They are particularly interested in how θ relates to the old treatment's success probability, believed to be about 0.2. They have reason to hope θ is considerably greater, perhaps 0.8 or even greater. To obtain evidence about θ, they carry out a study in which the new treatment is given to 17 subjects, and find that it is successful in nine.

How would an evidentialist test the composite hypothesis that the true proportion (θ) is greater than 0.2?

The maximum likelihood or ML 0.5294. The $k = 32$ support interval for quite strong evidence is [0.233, 0.811]. Royall would say that for any value θ' outside of this interval, there is quite strong evidence for the maximum likelihood estimate or MLE as opposed to θ'. For an evidentialist, this is sufficient to infer with strong evidence that (θ) > 0.2, even though the likelihood of the MLE is not the likelihood of the composite. The following considerations support this claim.

(1) There is quite strong evidence against any value outside the interval relative to a value inside the interval (i.e. the maximum likelihood estimate).
(2) No two values inside the interval can be quite strongly differentiated.
(3) (1) and (2) together imply that there is quite strong evidence that the true proportion θ is in the support interval [0.233, 0.811].
(4) Since 0.2 is entirely below the support interval, there is therefore quite strong evidence that the 0.2 is less than the true proportion.

It does make explicit that there are limits on how high the true proportion is likely to be.

We will use a notion called "the probability of misleading evidence" which will be discussed in much detail in Chap. 8. The probability for the presence of evidence for a hypothesis is called misleading because although there is probability for the presence of the evidence for the hypothesis, the latter is in fact false. If one had set up $k = 32$ (quite strong evidence) then the probability of misleading evidence for this statement is $M_G < 1/32 = 0.031$. The M_L represents the probability of misleading *local* evidence *after* the data have been gathered. The M_G represents the probability of misleading *global* evidence *before* the data have been gathered. Both M_L and M_G represent a

---

[53]This assumption is common to different schools of statistics. Both Royall (1997), who is committed to a likelihoodist (and not, as already noted, Bayesian) approach, and Mayo (1996) who operates within the error-statistical framework, also take the assumption for granted (at least Mayo did in 1996, although she does so no longer; again see Mayo 2014).

[54]Royall (1997, pp. 19–20).

bound on the probability of misleading evidence for a hypothesis. The post hoc probability of misleading evidence is a little lower MG = 0.011.

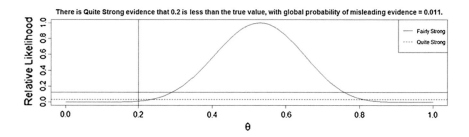

Taper and Lele (2011) suggest that since there is an estimated parameter in finding the MLE, the M_G is biased high, and that composite intervals should use a biased corrected estimate of the likelihood. We use Akaike's bias correction (for simplicity). With the bias correction, the quite strong evidence support interval is a little bit wider at [0.201, 0.840]. The inference is still the same. There is still quite strong evidence that the true proportion is greater than 0.2, but now the post hoc probability of misleading evidence is slightly greater at 0.030. Using the more severe Swartz bias correction, we find that there is only fairly strong evidence for 0.2 being less than the true value, with a M of 0.045 (see also Taper and Ponciano 2016).

References

Achinstein, P. (Ed.). (1983). *The concept of evidence*. Oxford: Oxford University Press.
Achinstein, P. (2001). *The book of evidence*. Oxford: Oxford University Press.
Bandyopadhyay, P., & Brittan, G. (2010). Two Dogmas of Strong Objective Bayesianism. *International Studies in the Philosophy of Science, 24*, 45–65.
Bandyopadhyay, P., & Forster, M. (Eds.). (2011). *Handbook of the philosophy of statistics*. Amsterdam, Elsevier: North-Holland.
Bandyopadhyay, P. (2007). Why Bayesianism? A Primer on a Probabilistic Theory of Science. In Upandhyay, U. Sing & D. Dey, (Eds.). *Bayesian statistics and its applications*. New Delhi: Anamaya Publishers.
Basu, A., Shioya, H., & Park, C. (2011). *Statistical inference: the minimum distance principle approach*. New York: Chapman and Hall (CRC Press).
Berger, J. (1985). *Statistical analysis and Bayesian analysis* (2nd ed.). New York: Springer.
Berkovitz, J. (2015). The Propensity Interpretation of Probability: A Re-evaluation. *Erkenntnis*, pre-publication version available at http://link.Springer.com/article/10.1007%2Fs10670-014-9716-8.
Carnap, R. (1950). *Logical foundations of probability*. Chicago: University of Chicago Press.
Crupi, V., Chator, N., Tenori. K. (2013). New Axioms for Probability and Likelihood Measures. *British Journal for the Philosophy of Science 64*, 189–204.
Earman, J. (1992) *Bayes or bust?* Cambridge, MA: MIT Press.
Edwards, M., Lindman, H., & Savage, L. (1963). Bayesian Statistical Inferences for Psychological Research. *Psychological Review, 70*(3), 193–242.

Eells, E. (2006). Confirmation Theory. In Sarkar and Pfeiffer, (Eds.). *The philosophy of science: A-M*. London: Taylor and Francis.

Eells, E., & Fitelson, B. (2000). Measuring Confirmation and Evidence. *Journal of Philosophy, 17*, 663–672.

Fitelson, B. (1999). The Plurality of Baeysian Measures of Confirmation and the Problem of Measure Sensitivity. *Philosophy of Science, PSA, 66*(1), 362–378.

Forster, M. & E. Sober. (2004). Why Likelihood. In (Taper and Lele, 2004).

Glymour, C. (1980). *Theory and evidence*. Princeton: Princeton University Press.

Hàjek, A. (1997). 'Mises Redux'—Redux: Fifteen Arguments Against Finite Frequentism. *Erkenntnis, 45*, 209–227.

Hitchcock, C., & Sober, E. (2004). Prediction versus Accommodation and the Risk of Over-fitting. *British Journal for the Philosophy of Science, 55*, 1–34.

Hawthorne, J. (2011) Confirmation. In (Bandyopadhyay and Forster, 2011).

Kaplan, M. (1996). *Decision theory as philosophy*. Cambridge: Cambridge University Press.

Kass, R., & Rafetry, A. (1995). Bayes Factors. *Journal of the American Statistical Association, 90* (430), 773–795.

Keynes, J. M. (1921). *A treatise on probability*. London: Macmillan.

Kotzen, M. (2012). Dragging and Confirming. *Philosophical Review* 121.1, 55–93.

Kyburg, H. (1961). *Probability and the logic of rational belief*. Wesleyan, CT: Wesleyan University Press.

Lele, S. (2004). Evidence Function and the Optimality of the Law of Likelihood. In (Taper and Lele, 2004).

Levi, I. (1967). Probability Kinematics. *British Journal for the Philosophy of Science, 18*, 200–205.

Lewis, D. (1980). A Subjectivist's Guide to Objective Chance. In R. C. Jeffrey (Ed.), *Studies in inductive logic and probability* (Vol. 11). Berkeley: University of California Press.

Lindsay, B. (2004). Statistical Distances as Loss Functions in Assessing Model Adequacy. In (Taper and Lele, 2004).

Mayo, D. (1996). *Error and the growth of experimental knowledge*. Chicago: University of Chicago Press.

Mayo, D. (2014). How likelihoodists exaggerate evidence from statistical tests. https://errorstatistics.files.wordpress.com/2014/11/images2.jpeg.

Popper, K. (1959) (first published in 1938). *The logic of scientific discovery*. New York: Basic Books.

Popper, K. (1957). The Propensity Interpretation of the Calculus of Probability and the Quantum Theory. In: Körner, S., (Ed.) *Observation and interpretation: proceedings of the ninth symposium of the colston research society, university of bristol*, pp. 65–70 and 88–89.

Reichenbach, H. (1949). *The theory of probability*. Berkeley: University of California Press.

Royall, R. (1997). *Statistical evidence: a likelihood paradigm*. New York: Chapman Hall.

Royall, R. (2004). "The Likelihood Paradigm for Statistical Evidence." In (Taper and Lele, 2004).

Salmon, W. (1983). Confirmation and Evidence. In (Achinstein, 1983).

Shafer, G. (1976). *A mathematical theory of evidence*. Princeton: Princeton University Press.

Silverman, B. (1986). *Density estimation for statistics and data analysis*. London: Chapman and Hall/CRC Press.

Skyrms, B. (1986). *Choice and chance* (3rd ed.). Belmont, CA: Wadsworth Publishing.

Sober, E. (1993). Epistemology for Empiricists. In French, Uehling, & Weinstein (Eds.), *Midwest studies in philosophy* XVIII: 39–61.

Taper, M., & Lele, S. (Eds.). (2004). *The nature of scientific evidence*. Chicago: University of Chicago Press.

Taper, M., & S. Lele. (2011). Evidence, Evidence Functions, and Error-Probabilities. In (Bandyopadhyay and Forster, 2011).

Taper, M., & J. Ponciano. (2016). Evidential statistics as a statistical modern synthesis to support 21st century science. *Population Ecology, 58*, 9–29.

Taper, M., Staples, D., & Shephard, B. (2008). Model Structure Adequacy Analysis: Selecting Models on the Basis of Their Ability to Answer Scientific Questions. *Synthèse, 163*, 357–370.

Von Mises, R. (1957). *Probability, statistics, and truth, revised* (English ed.). New York: Macmillan.

Williamson, T. (2000). *Knowledge and its limits*. Oxford: Oxford University Press.

Chapter 3
Confirmation and Evidence Distinguished

Abstract It can be demonstrated in a very straightforward way that confirmation and evidence as spelled out by us can vary from one case to the next, that is, a hypothesis may be weakly confirmed and yet the evidence for it can be strong, and conversely, the evidence may be weak and the confirmation strong. At first glance, this seems puzzling; the puzzlement disappears once it is understood that confirmation is of single hypotheses, in which there is an initial degree of belief which is adjusted up or down as data accumulate, whereas evidence always has to do with a comparison of one hypothesis against another with respect to the data and is belief-independent. Confusing them is, we suggest, a plausible source of the so-called "base-rate fallacy" identified by Kahneman and Tversky which leads most of us to make mistaken statistical inferences. It is also in the background, or so we argue in some detail, of the important policy controversies concerning human-induced global warming.

Keywords Degree-of-belief · Likelihood ratio · Diagnostic testing · Base-rate fallacy · Global warming hypothesis · Uncertain inference

Confirmation as Increased Degree-of-Belief, Evidence as a Likelihood Ratio > 1

The notion of confirmation relates to single hypotheses. From a Bayesian perspective, it has to do with the ways in which, and the degree to which, belief in a hypothesis is reasonable; the degree to which belief in a hypothesis H is reasonable is a function of the degree to which data D confirm it. In general, D confirm H just in case $\Pr(H \mid D) > \Pr(H)$, that is, the data raise its posterior relative to its prior probability. The degree to which its posterior probability has been raised or lowered is, in turn, a function of the prior probability of the hypothesis, the probability of the data on the hypothesis, and what is sometimes called the "expectedness" or marginal probability of observing the data averaged over both hypotheses H and $\neg H$ (in the

© The Author(s) 2016
P.S. Bandyopadhyay et al., *Belief, Evidence, and Uncertainty*,
Philosophy of Science, DOI 10.1007/978-3-319-27772-1_3

simplest case). It follows that the degree to which we are justified in believing that a hypothesis is true presupposes no inter-hypothetical comparison.[1] Whether data constitute evidence, on the other hand, has to do with the ways in which they serve to distinguish and compare competing hypotheses. It is a three-part relation involving data and two hypotheses. Data that cannot tell for or against such hypotheses do not constitute evidence for one or the other. A natural way to express this very basic intuition is through the use of likelihood ratios. Thus, data D constitute (positive) evidence for hypothesis H just in case the ratio of likelihoods, $Pr(D \mid H)/Pr(D \mid H')$ is greater than 1, where H and H' are not necessarily mutually exclusive. If D is equally likely on H and its competitors, then D does not constitute evidence for any of them. The point needs to be emphasized: data by themselves do not constitute evidence, but only in a context provided by competing hypotheses. The same data can constitute evidence for a hypothesis in one context and not constitute evidence for it in another context, depending on the alternative hypotheses with which it is compared.

Confirmation and evidence as just characterized are two very different notions. Degree of confirmation as we understand it here[2] is the difference between posterior and prior probabilities; it must range from anywhere above 0 to anywhere below 1. Evidence as a ratio between likelihoods can range between 0 and ∞ as its limits. Moreover, there is a coherence condition on confirmation that need not be satisfied by our account of evidence: if H_1 entails H_2, then the probability of H_1 cannot exceed the probability of H_2. In addition, the notion of justification is agent-sensitive; it depends on a distribution of prior probabilities on hypotheses, and relates, like belief generally, to what is in one's head. The notion of evidence is agent-independent; it depends on a ratio of likelihoods already determined, and to this extent has to do with how things stand in the world, independent of the agent's belief or knowledge. Finally, although evidence is accompanied by confirmation and vice versa when the hypotheses being compared are mutually exclusive and jointly exhaustive, even then the relation is not linear. Indeed, in sample cases they can vary dramatically. A hypothesis for which the evidence is very strong may not be very well confirmed, a claim that is very well confirmed may have no more than weak evidence going for it.

---

[1]That is, we do not need to know the posterior probabilities of the mutually exclusive and jointly exhaustive H's and not-H's in order to calculate the posterior probability of H in the simple cases that we use to illustrate our point. It might be thought that there is a cryptic reference to two hypotheses in the determination of $Pr(D)$ in the denominator of Bayes Theorem, since it involves averaging the data over H and ~H. But there is only one hypothesis, H, and the data are averaged over whether it is true or false. From this point of view, the expression "mutually exclusive and jointly exhaustive hypotheses" is misleading. Asserting "~H" (is true) is just another way of saying that "H" is false. In more complex cases, we do need to know the priors of all of the hypotheses being considered in order to calculate the marginal probability of the data.

[2]Since it is the Bayesian measure most frequently encountered in the literature. As noted in Chap. 2, there are other measures of confirmation and evidence than those we are taking as paradigm.

A Diagnostic Example: Testing for Tuberculosis

On the basis of a very extensive sample, viz., 100,000 people, the probability of having a positive X-ray for those people infected with TB was near 0.7333, and the probability of a positive X-ray for those people not similarly infected is near 0.0285. Denote this background information regarding such probabilities as B. As earlier, let H_1 represent the hypothesis that an individual is suffering from tuberculosis and $\sim H_1$ the hypothesis that she is not. These two hypotheses are clearly mutually exclusive and jointly exhaustive. Finally, assume that D represents a positive X-ray test result.

The task is to find $Pr(H_1 \mid D \;\&\; B)$, the posterior probability that an individual who tests positive for tuberculosis actually has the disease. Bayes theorem enables us to obtain that probability. In order to apply the theorem, however, we first need to know $Pr(H_1)$, $Pr(\sim H_2)$, $Pr(D \mid H_1 \;\&\; B)$ and $Pr(D \mid \sim H_1 \;\&\; B)$. $Pr(H_1)$ is the prior probability that an individual in the general population has tuberculosis. Because the individuals in different studies who showed up in the medical records were not chosen from the population at random, the correct frequency-based prior probability of the hypothesis could not be inferred from the large data-set referred to above. Yet in a 1987 survey (Pagano and Gauvreau 2000),[3] there were 9.3 cases of tuberculosis per 100,000 people. Consequently, $Pr(H_1) = 0.000093$. Hence, $Pr(\sim H_1) = 0.999907$. As already indicated, on the basis of a large data-set kept as medical records, we may take the following probabilities at face-value: $Pr(D \mid H_1 \;\&\; B)$, the probability of a positive X-ray given that an individual has tuberculosis, $= 0.7333$; $Pr(D \mid \sim H_1 \;\&\; B)$, the probability of a positive X-ray given that a person does not have tuberculosis, $= 1-Pr(\sim D \mid \sim H_1 \;\&\; B) = 1-0.9715 = 0.0285$. Using all of this information, we compute $Pr(H_1 \mid D \;\&\; B) = 0.00293$.

Because most people do not have tuberculosis, the surprising result is that, even though the test has fairly good specificity, most positive tests are false positives. For every 100,000 positive X-rays, only 239 signal genuine cases of tuberculosis. For a test of an individual, the posterior probability is very low, although it is slightly higher than the prior probability. The posterior-prior difference, our measure of confirmation, is only 0.00284. The hypothesis is not very well confirmed, yet at the same time, the LR, viz., 0.7333/0.0285 (i.e., $Pr(D \mid H_1 \;\&\; B)/Pr(D \mid \sim H_1 \;\&\; B)$ $= 25.7$, is high. Therefore, the test for tuberculosis has a great deal of evidential significance.[4] A crucial aspect of our account of evidence is that a datum could be evidence for a hypothesis as against its alternative independent of whether the agent believes/knows that the datum has been gathered or the hypothesis is true. The TB example shows that the positive test, D, is strong evidence that the individual is more likely (approximately 26 times more likely) to have the disease than not, independent of whether D or H_1 or $\sim H$ is believed or known to be true.

---

[3]Subsequent research shows that this frequency-based prior probability still holds for the US population.

[4]Recall that a standard benchmark for "strong" evidence is an LR > 8.

Appraising the Human-Induced Global Warming Hypothesis

A much more complex example of widespread failure to make and appreciate the distinction between confirmation and evidence in our terms leads to overlooking an element in a controversy with sweeping social, political, and economic dimensions. The controversy has to do with the anthropogenic "global warming" hypothesis, that is, the hypothesis that present warming trends are human-induced. A wide variety of data raise the probability of the hypothesis, in which case they confirm it. Indeed, in the view of most climatologists, this probability is very high. The Intergovernmental Panel on Climate Change contends that most of the observed temperature increase since the middle of the 20th century was caused by increasing concentrations of greenhouse gases resulting from human activity such as fossil fuel burning and deforestation.[5] In part this is because the reasonable prior probability that global warming is human induced is very high, as against the rather low prior in the tuberculosis example. In the TB example, the prior was assigned on the basis of relative frequencies. In the global warming case, it is assigned not on the basis of relative frequencies (which for better or worse are unavailable), but because there is a very good theoretical explanation linking human activity to the "greenhouse effect," and thence to rising temperatures. In part, the posterior probability of the hypothesis that global warming is human-induced is even higher because there are many strong correlations in the data, at least for the last 650,000 years or so (the period for which some reliable data have been extracted from ice cores), between the burning of fossil fuels, in particular, and rising globally-averaged temperatures, most especially since the Industrial Revolution took hold in the middle of the 18th century.[6] Not only is there a strong hypothesized mechanism for relating green-house gases to global warming, this mechanism has been validated in detail by physical chemistry experiments on a micro scale, and as already indicated there is a manifold correlation history between estimated CO_2 levels and estimated global temperatures. Of course, some climate skeptics emphasize how difficult it is to get standardized and reliable data for such a long period of time and from so many different places, others point out that it has not always been true that changes in CO_2 levels precede changes in temperature,[7] still others draw attention to the apparent but controversial datum that global temperatures have not been rising for the last 15 years or so,[8] despite the fact that atmospheric CO_2 levels continue to do

---

[5]See IPCC (2007).

[6]See Spahni et al. (2005), Siegenthaler et al. (2005), and Petit et al. (1999).

[7]Although over the past 200 years, when a rise of temperatures and the vastly increased use of fossil fuels both occurred, the rise in CO_2 levels invariably preceded the temperature rise.

[8]2014, however, has apparently been the hottest year since accurate temperature records began to be kept. This claim has been disputed by the well-known Harvard astrophysicist Willie Soon, but taking the observation at face value, the evidence for a lack of recent temperature increase is greatly weakened. Preliminary results for 2015 indicate that it was hotter still.

so.[9] But the main skeptical lines of argument mounted by such credible climate skeptics as Richard Lindzen and Freeman Dyson (who, although a celebrated physicist, and not a climatologist, was in the vanguard of serious students of climate change in the 1970's) are that (a) the likelihood of the data on the alternative default (certainly simpler) hypothesis, that past and present warming is part of an otherwise "natural" long-term trend and therefore not (significantly) human-induced or "anthropogenic," is just as great,[10] (b) that the data are at least as likely on other, very different hypotheses,[11] among which solar radiation and volcanic eruption[12] (c) that not enough alternative hypotheses have been considered to account for the data.[13] That is, among credible climate skeptics there is some willingness to concede that burning fossil fuels leads to CO_2 accumulation in the atmosphere and that carbon dioxide is a greenhouse gas that traps heat before it can escape into the atmosphere,[14] and that there are some data correlating a rise in surface temperatures with CO_2 accumulation. But, the skeptics continue, these correlations do not "support," still less "prove," the anthropogenic hypothesis because they can be equally well accounted for on the default, "natural variation" hypothesis or by some specific alternative. In our terms, there is no or only very weak *evidence* for the anthropogenic hypothesis. In their view, alternative hypotheses are not taken seriously enough and therefore are not systematically and rigorously compared with it. Since there is little evidence for the hypothesis, it is not, the skeptics conclude, very well confirmed (and for this and other reasons massive efforts to reduce carbon emissions are a costly mistake). But this conclusion rests on a conflation of evidence

---

[9]There has been a great deal of controversy about the nature of the data and the ways in which correlations between CO_2 levels and temperature changes are established. One among many criticisms of both data and correlations taken to support the human-induced warming hypothesis is in Douglass and Christy (2009). For a history of the data/correlation controversy by the person who was chiefly responsible for developing the initial data sources and correlation models, see Mann (2012). Mann's methodology and his models have been revised, extended, but in the main confirmed by a number of subsequent studies, including (Marcott et al. 2013), which uses marine and terrestrial fossils, inter alia, from ocean and lakebed sediments, as well as ice cores and tree-ring data (which don't extend over the entire period).

[10]See Lindzen (2010) who asserts that IPCC computer models don't "accurately include any alternative sources of warming—most notably, the natural, unforced variability associated with phenomena like El Niño, the Pacific Decadel Oscillation, etc".

[11]Lindzen argues not so much that there are better alternative hypotheses but that the anthropogenic hypothesis incorporates assumptions about negative radiation feedback which have not been tested against their positive feedback rivals, i.e., the global thesis has indispensable components for which there is as yet no "evidence." See Lindzen and Y-K Choi (2009). Lindzen's argument has been much criticized. See, e.g., (Lin et al. 2010).

[12]See Camp and Tung (2007) and Rypdal (2012).

[13]See, in particular, Dyson (2009).

[14]According to Christy (2014), these are the "two fundamental facts" that everyone must accept. Christy, a credible global warming skeptic, is, however, very dubious about the claim that there are strong correlations between a rise in surface temperatures and CO_2 accumulation and critical of the way in which "surface temperatures" have been defined and measured.

with confirmation, and provides a striking reason why it is necessary to distinguish them.

Data are evidentially significant only if they discriminate between hypotheses, and such data in the case of human-induced global warming have been difficult to come by. That fact has premised at least part of the sceptics' argument. But such data have increasingly been identified.

We think, in fact, that variations on the anthropogenic hypothesis are both well confirmed by the data and supported by the evidence. In implicitly denying the second claim, that the anthropogenic hypothesis is supported by evidence in our precise sense of the word, some skeptics claim that the rise in atmospheric CO_2 comes from the ocean, and not from burning fossil fuel. Its rise is therefore "natural,"[15] or at the very least it is as likely that the greenhouse-gases responsible for temperature rise come from the ocean, for example, as it is that they are produced as a result of human activities.

But a crucial observation has been made to distinguish the two sub-hypotheses, H_1, that the CO_2 increases responsible for (longer-term) temperature rise come from burning fossil fuel, and H_2, that the ocean is responsible. Most carbon atoms have an atomic mass of 12, but about 1% have an atomic mass of 13. Both kinds can form CO_2 molecules, $^{12}CO_2$ and $^{13}CO_2$, distinguishable in the laboratory. To put a somewhat complex story very simply,[16] it can be shown that if the CO_2 atmosphere comes from the surface (and not the depths) of the ocean, then $^{13}CO_2$ will increase over time. If the CO_2 comes from fossil fuel burning, then the relative abundance of $^{13}CO_2$ to $^{12}CO_2$ will decrease. Experimental results show that the $^{13}CO_2/^{12}CO_2$ ratio is decreasing,[17] evidence for the hypothesis that fossil fuels rather than surface water is mainly responsible for rising levels of CO_2 in the atmosphere, and hence (on the assumption that rising levels of CO_2 are a cause of rising temperatures) for the anthropogenic hypothesis.

It is not our aim to show that the anthropogenic hypothesis is (very likely) true, but to indicate that front-page controversies as well as otherwise obscure journal articles have the conflation between evidence and justification/confirmation as one of their sources,[18] and to issue the methodological advice that it is never enough in

---

[15]What follows draws from a very accessible overview by a Professor in the Department of Physics and Astronomy at the University of Las Vegas (Farley 2008).

[16]Mark Greenwood and his colleagues have, to give but one example, shown just how complex the application of statistical techniques to the detection of changes that provide support for the global warming hypothesis is. See Greenwood et al. (2011).

[17]IPCC (2004, Chap. 2, page 138).

[18]Another source of the misguided controversy has to do with the Bayesian account of confirmation that we have taken as paradigm. The prior probability of the hypothesis that global warming is not human-induced, is admittedly subjective, and for many people its rough determination depends not only on the plausibility of the "greenhouse gas" model, but on such otherwise extraneous considerations as that attempts to limit fossil fuel emissions will damage the economy and lead to the loss of jobs. Expected economic consequences often bleed over into the generation of priors concerning the existence and causes of global climate change, neither of which are in themselves economic hypotheses.

testing a hypothesis to gather data that confirm or disconfirm it. One must also find ways in which to discriminate and judge the hypothesis against its rivals with respect to the data. This is the challenge that is now being met by climate scientists, to provide evidence for, in addition to confirmation of, the anthropogenic hypothesis.

A Possible Source of the Base-Rate Fallacy

The confirmation/evidence distinction has a number of significant corollaries. One of them concerns a widespread inferential error that has been much discussed over the past generation or so, A key premise in our simple schematic demonstration that it is possible to have very strong evidence and yet a very low degree of confirmation is that for every 100,000 positive X-rays, only 239 signal genuine cases of tuberculosis. Yet the premise leads to a result that strikes many people as wildly counter-intuitive. It is therefore worth our while to examine the intuition. It rests on what has come to be called "the base-rate fallacy." The base-rate fallacy ascribes the propensity that the general populace has to miss-estimate the probability of complex events to an overweighting of the frequency of secondary attributes to the frequency of the fundamental or base event in the overall population. We speculate that the fallacy stems from our common tendency to conflate evidence and confirmation. Further, we take the near-ubiquity of the base-rate fallacy as "evidence," so to speak, that a distinction between them should be made.

As we underlined at the outset, uncertainty is an inescapable aspect of human life. For one thing, it makes the analysis of scientific inference very difficult. For another thing, and more importantly, it forces us to make significant choices on the basis of inevitably partial information. Amos Tversky and Daniel Kahneman have argued famously[19] that we often make mistakes when we reason probabilistically, as we must, on the basis of such partial and uncertain information. They set out case after case of probabilistic inferential error. Perhaps the most wide-spread of these errors is the base-rate fallacy. It can be illustrated using a variant of the TB example just discussed.

There are a couple of ways the base-rate fallacy has gained currency.[20] Here, we follow David Papineau's characterization of the base-rate fallacy (Papineau 2003). Groups of subjects are given a case and a multiple-choice question to answer with respect to it:

---

[19]See their classic paper, "Judgment Under Uncertainty: Heuristics and Biases," re-published as Appendix A in Kahneman (2011).

[20]In the original question in the Harvard Medical School Test the subjects were told that nothing was known about the person's symptoms or background, justifying the use of the base-rate prior, and they were asked to give the probability, on the assumption, that the person had the disease (see Howson (2000) for this discussion). We are thankful to Howson for some clarification here.

Someone is worried that she has tuberculosis. 1 percent of the population has tuberculosis. There is a simple and effective test, which identifies the presence of tuberculosis in everyone who has it, and only gives a false positive result in 10 percent of the people who do not have it. She takes the test, and gets a positive result. What is now the closest probability that she in fact has tuberculosis?

(A) 90 percent
(B) 10 percent
(C) 50 percent
(D) 89 percent

Again and again, the subjects fail to give the correct response. Their average estimate of the probability that she suffers from the disease is 85 %, whereas the correct answer (computed using Bayes Theorem as we have done above) is about 10 %. Why do they fail to get it? According to Kahneman and Tversky, they fail because they overlook the base-rate of the disease in the population; as we noted, no more than roughly 1 % of the population suffers from the disease. In a variety of uncertain situations, we can safely ignore the base-rate. This leads us, on the Kahneman and Tversky "heuristics and biases" approach, to ignore it generally. They explain why humans generally get the wrong estimate by arguing that human beings adopt certain heuristic strategies in solving problems, strategies that generally provide useful short-cuts to reasonably accurate answers, but that also bias subjects irrationally toward certain kinds of mistakes.[21]

In our view, the distinction between confirmation and evidence provides an alternative, equally plausible, rationale for the subjects' failure to get the TB probability right. The base-rate fallacy results from the subjects' (certainly not the investigators') conflation of the two notions. They think the evidence question has been asked, whereas in fact the confirmation or belief question was asked. That is, given the data about the likelihood of positive and negative test results on the hypotheses that the people tested were and were not affected, they rightly concluded that the data provided very strong evidence that a person who tested positive was much more likely to have rather than be free of TB. Given the data, the evidence on our LR account of it provide 10 times more support for the person in question having the disease than not having it.[22] This is *strong* evidence for the hypothesis that she has the disease, despite the fact that the confirmation of it is low. So our diagnosis of the fact that people tend in certain kinds of cases to ignore the base-rate is that they mistakenly take the question, "what is the probability post-test that she

---

[21]We have argued that another heuristic short-cut, using the so-called "collapsibility principle" across the board, results in the celebrated Simpson Paradox (Bandyopadhyay et al. 2011) and have carried out our own experiments to confirm it, an exercise in "experimental philosophy".
[22]$\Pr(D \mid H)/\Pr(D \mid \neg H) = 1/0.1 = 10$.

has tuberculosis?" as a question about the strength of the evidence, and not about the degree of confirmation that the data give to the hypothesis that she has the disease, which is only 10 %.[23]

Provisional Conclusion

We have demonstrated using the difference between the posterior and prior and the likelihood ratio measures of confirmation and evidence respectively that there are cases in which the evidence is strong and the confirmation limited. There are similar cases in which the evidence is weak and the confirmation is high, and there are also tests the results of which provide neither strong evidence nor high confirmation. Those scientific hypotheses for which the data provide both strong evidence and high confirmation are, and should be, commonly accepted, at least until such time as new hypotheses out-compete them with respect to the data.[24]

However, the demonstration is not restricted to these two measures. The same result can be found whenever the confirmation measure depends on the prior and the evidence measure does not. For instance, we have also found this feature using other well-known confirmation measures, including r and S.[25] Evidence and confirmation occasionally increase together, but not to the point where strong evidence or high confirmation entail the other. They are distinct, even to the point where evidence sometimes supports a rival, and yet the degree of confirmation of the

---

[23]Referring to the Harvard Medical School Test, subjective Bayesians might contend that the answer the subjects gave were definitely wrong, and *not just misunderstanding* of what was required. They might add that in our version the person is already worried that they have TB, and so the prior is already much larger than the base rate. We will make two comments here. First, we agree with subjective Bayesians that subjects committed a probability error regarding the base-rate in our example. However, as contrasted with subjective Bayesians, we are able to provide an explanation for why subjects might have committed that sort of error. Second, if what we have stated is not the exact formulation of the base-rate fallacy typically understood then we need to know the entire probabilistic machinery at least conceptually required to address the way we discussed the fallacy because of the claim made by the subjective Bayesian. The onus is on the subjective Bayesian to offer that probabilistic mechanism regarding how the issue at stake can be handled within subjective Bayesianism.

[24]An anonymous referee of a paper containing the tuberculosis example claims that "the measures of confirmation literature…already provide the conceptual resources to acknowledge cases in which, by some measures, a hypothesis is greatly confirmed by a piece of evidence (e.g., when it ends up 100x as probable as it previously was), even though it does not end up highly probable nor does its probability change a great deal in absolute terms. This is what's going on in the authors' discussion of the TB example…" But it should be clear that "greatly confirming by a piece of evidence" (*sic*) is not at all tantamount to having strong evidence that the hypothesis is much better supported by the data than its rivals. In the TB case, the likelihood of a positive result on the hypothesis that the person tested has TB is much greater than on the hypothesis that she does not, independent of whether the first hypothesis is "greatly confirmed" by the data. As the referee's criticism indicates, it is very difficult to shake the conflation of "evidence" with "data".

[25]See Fitelson (2001).

target hypothesis is nonetheless high.[26] In the case of such well-known diagnostic tests as the PAP Smear for cervical cancer (to be worked out in Chap. 6), the data do no more than provide weak evidence and do not raise the posterior probability appreciably, whereas in paradigm episodes in the history of science they do both.[27]

Uncertain Inference

The premise with which we began this monograph is that a great deal of scientific as well as day-to-day inference is uncertain; many of the conclusions drawn go beyond the data on which they are presumably based in the twin sense that the data gathered can be correct and the conclusion false and that adding or subtracting premises from the arguments in which they figure can undermine or strengthen support for the conclusions, in which case the reasoning is not monotonic. In both kinds of cases, the inference from data-premises to conclusion is not deductively valid, i.e., the truth of the premises does not guarantee the truth of the conclusion.

To mark the contrast, uncertain inference is usually called "inductive."

A variety of questions are raised by uncertain or inductive inference. Three have been the focus of philosophers' attention. First, "when and under what conditions does evidence confirm a hypothesis?" To answer this question is to provide a list of criteria. Second, "what rules, if any, will (if followed) make reasonably sure that the hypothesis is well-supported by a given set of data-premises and perhaps also to what degree?" To answer this question is to provide a "logic" of inductive reasoning. Third, "how can the choice of one set of rules as against another be rationally justified?" To answer this question is to solve/dissolve at least one form of Hume's notorious "problem of induction."

In our view, all three questions are defective. That the third question has not been answered, despite enormous efforts to do so, in a generally accepted way provides an inductive and ironic argument for the claim that it is seriously flawed. A number of philosophers have tried to pinpoint the flaws, beginning with Hume, the first person in the western world to ask the question in something like this form. This is not the place to canvas their attempts. As for the second question, there is no doubt that at least some of the rules of inductive inference proposed both describe

---

[26]With multiple hypotheses it is easy to create scenarios where one model's posterior probability is greater than its prior, in which case the model is confirmed, but have evidence against it. Consider three urns (A, B, and C) each containing 100 black or white balls. Urn A has one black ball and 99 white balls, Urn B has two black balls, and Urn C has no white balls. You are presented with an urn, but don't know which it is. For whatever reason, you strongly believe that it is Urn C, but allow for the possibility that it could be A or B by assigning prior probabilities of 0.01, 0.1, and 0.89 to A, B, and C respectively. You draw a white ball randomly from the unidentified urn. Urn B is strongly confirmed because its posterior probability of 0.90 is much greater than its prior; however, there is weak evidence *against* B relative to A in that the LR B/A is 0.99.

[27]Bandyopadhyay and Brittan (2006).

certain aspects of successful scientific practice and incorporate deep intuitions that render hypotheses more or less probable. We have discussed several of these rules in some detail. But a key implication of our distinction between confirmation and evidence is that there is no one set of rules that will, if followed, make reasonably sure that a hypothesis is supported by a given set of data premises.[28] There is no "logic" (on the model of deductive logic) of inductive reasoning. For one thing, what inferential procedures are "best" depends on the specific questions the investigator wants to answer and on the aims of the investigation. For another thing, inductive inference is "uncertain" in a variety of ways. As we noted at the outset, the conclusion of such an inference can never be more than merely probable or likely. But the inferences are also made uncertain by the fact that a choice, which itself is not rule-governed, has to be made among possible procedures to be used and, as we will see at greater length in Chap. 8, even when a procedure has been chosen, the ever-present possibility of "misleading evidence" makes its application itself uncertain.

But the main argument against a "logic of inductive inference," and the focus of this monograph, is that the premise of the first question, "when and under what conditions does evidence confirm a hypothesis?" is mistaken and therefore badly misleading. "Evidence" does not "confirm." The two concepts provide very different accounts of how data bear on the estimation of parameters and the choice of hypotheses or models. It is this conflation, we suggest, that has given rise to the "which set of rules is best?" and "how can they be justified?" questions, and in the process set philosophers and philosophically-minded statisticians off on what has been an enlightening and in many ways productive, but ultimately unsuccessful quest.

In the next chapter, we address some of the traditional criticisms made of the components of our bifurcated account and further clarify our intentions.

References

Bandyopadhyay, P., & Brittan, G. (2006). Acceptance, evidence, and severity. *Synthèse, 148*, 259–293.
Bandyopadhyay, P., Nelson, D., Greenwood, M., Brittan, G., & Berwald, J. (2011). *The Logic of Simpson's Paradox. Synthèse, 181*, 185–208.
Christy, J., (2014, February 19). Why Kerry is flat wrong on climate change. *Wall Street Journal.*
Camp, C., & Tung, Ka Kit. (2007). Surface warming by the solar cycle as revealed by the composite mean difference projection. *Geophysical Research Letters, 34*, L14703.
Dyson, F. (2009). Interview, *Yale Environment 360.*
Douglass, D., & Christy, J. (2009). Limits on CO_2 climate forcing from recent temperature data of earth. *Energy and Environment, 20*(1–2), 177–189.
Farley, J. (2008). The scientific case for modern anthropogenic global warming. *Monthly Review, 60*, 3.

---

[28]The following draws upon and is made more precise in Lele (2004).

Fitelson, B. (2001). A Bayesian account of independent evidence with applications. *Philosophy of Science, PSA, 68*(1), 123–140.

Greenwood, M., Harwood, C., Moore, D. (2011). In (P.S. Bandyopadhyay, M. Forster (eds.)).

Howson, C. (2000). *Hume's problem: Induction and the justification of belief.* Oxford: Clarendon Press.

IPCC (Intergovernmental Panel on Climate Change). (2004). *IPCC 2004 Report.*

IPCC. (2007). *Climate change 2007: The physical basis, contribution of working group I to the fourth assessment report of the IPCC.*

Kahneman, D. (2011). *Thinking, fast and slow.* New York: Farrar, Stauss, and Giroux.

Kahneman, D., Slovic, P., & Tversky, A. (1982). *Judgment under uncertainty: Heuristics and biases.* Cambridge: Cambridge University Press.

Lele, S. (2004). Evidence Function and the Optimality of the Law of Likelihood. In M. Taper,S. Lele, (eds.) *The nature of scientific evidence.* 2004. Chicago: University of Chicago Press.

Lin, B., et al. (2010). Estimations of climate sensitivity based on top-of-atmosphere radiation imbalance. *Atmospheric Chemistry and Physics, 2*(19/2010), 1923–1930.

Lindzen, R. (2010, April 22). Climate science in denial. *Wall Street Journal.*

Lindzen, R., Choi, Y-K. (2009). On the determination of climate feedbacks from ERBE data. *Geophysical Research Letters* 36.

Mann, M. (2012). *The hockey stick and the climate wars.* New York: Columbia University Press.

Marcott, S., Shakun, J., Clark, P., & Mix, A. (2013). A reconstruction of regional and global temperatures for the Past 11.300 Years. *Science, 339*, 1198–1201.

Pagano, M., & Gauvreau, K. (2000). *Principles of biostatistics.* Garden Grove, CA: Duxbury Press.

Papineau, D. (2003): *The Roots of Reason*: Clarendon Press: Oxford.

Petit, J., et al. (1999). Climate and Atmospheric History of the past 420,000 Years from the Vostok Ice Core, Antarctica. *Nature, 406*, 695–699.

Rypdal, K. (2012). Global temperature response to radioactive forcing: Solar cycle versus volcanic Eruptions. *Journal of Geophysical Research* 117.

Siegenthaler, U., et al. (2005). Stable carbon cycle-climate relationship during the late pleistocene. *Science, 310*, 1313–1317.

Spahni, R., et al. (2005). Atmospheric methane and nitrous oxide of the late pleistocene from Antarctic ice cores. *Science, 310*, 1317–1321.

Chapter 4
Initial Difficulties Dispelled

Abstract In our view, data confirm a hypothesis just in case they increase its probability; they constitute evidence for one hypothesis vis-à-vis others just in case they are more probable on it than on its available rivals. In subsequent chapters, we go on to clarify and amplify the confirmation/evidence distinction. Before doing so, however, we need to consider various objections that might be made, not to the distinction itself but to the way in which we have formulated its principal elements. Four of these objections are standard in the literature. The first, third, and fourth raise questions concerning our analyses of both confirmation and evidence; the second has to do more narrowly with the application of Bayesian methods. Each suggests a different way in which our intentions in this monograph might be misunderstood.

Keywords Theory acceptance · Probabilistic measures · The "simple rule" · "Certain" data

Philosophical Analysis Cannot Deal with the Complexity of Theory Acceptance

The first objection is that a precise and formal account of evidence, and by the same token of confirmation, cannot be given.[1] It can be expressed in several ways. The main line of argument goes like this. If a precise account of either evidence or confirmation could be given, then debates between the relative merits of particular hypotheses, say between the advocates of the phlogiston theory of combustion and Lavoisier and his followers' oxygen theory, could be settled quickly. But they cannot be settled quickly, as one key episode after another in the history of science

---

[1]See, for example, Kitcher (2001, pp. 29–41), whose discussion itself runs together the two concepts we distinguish. "We lack an analysis that will reveal exactly which claims are justified (to what degree) by the evidence available at various stages" (p. 30). The distinction we make implies that no such analysis is possible. Kitcher does not himself advance the view that scientific claims cannot reasonably be appraised, but his discussion is so seamless and subtle that it is often difficult to know which side of the debate he is on.

© The Author(s) 2016
P.S. Bandyopadhyay et al., *Belief, Evidence, and Uncertainty*,
Philosophy of Science, DOI 10.1007/978-3-319-27772-1_4

demonstrates. Therefore, a precise account of neither can be given. The acceptance of scientific hypotheses is simply too messy an affair, depending as it does on adjusting philosophical constraints and shifting strategic considerations, not to mention cultural and social conditions, for the kind of artificial order we are imposing to have much effect.

Two subsidiary lines of argument are often invoked in this connection. One, famously ascribed to Thomas Kuhn,[2] is that the concepts of evidence and confirmation (along with such other ostensibly meta-theoretical concepts as explanation and observation) are themselves internal to particular scientific paradigms, and as such are subject to the shifts that such paradigms undergo. "The" concepts of evidence and confirmation are social-historical artifacts, tied to particular historical developments and valid only within individual communities of belief.

The other subsidiary line of argument is that no precise account of evidence or confirmation yet given has been immune to counter-examples, drawn either from the history of science or from the fertile imagination of philosophers and statisticians. Whatever the account, the latter have always been able to devise otherwise spurious hypotheses which are supported in all and only the same ways by the data for them as more apparently legitimate, even accepted, hypotheses, or on which the data are equally probable.

We will discuss the second subsidiary line of argument in the next section of this chapter and in Chaps. 9 and 11. It will be one of our main contentions that the "idle hypotheses"—the grue hypothesis, the dreaming hypothesis (that we live in a dream-world), and at least some of the rest—which are often invoked as counter-examples derive their intuition-pumping power from conflating the concepts of evidence and confirmation, and not from the attempt to make particular conceptions of them more precise.

So far as the Kuhnian line of argument is concerned, it does not much matter whether the concepts of evidence and confirmation are intra- or extra-paradigmatic. Our problems are not connected, at least not very directly, with paradigm shifts and historical considerations; they have been with us at least since the twin advent of modern science and philosophical reflection on it in the 16th and 17th centuries. Besides, precision, even at the expense of some simplification, has its own virtues. Clarification is one of them. Explanation is another. Our distinction is a kind of clarifying hypothesis which will explain why many philosophers, scientists and statisticians make certain kinds of mistakes. Even momentary stays against confusion are useful.

The more general theme, that if precise accounts of either evidence or confirmation could be given then scientific progress would have been much more orderly and disciplined than it in fact has been has been voiced by many philosophers over the past generation or so, particularly when it is conjoined with the call for a "naturalistic" epistemology, the replacing of "arm-chair" and normative accounts of how scientists do and should proceed by a scientific inquiry "describing processes

---

[2]Kuhn (1962/1970/1996).

that are reliable, in the sense that they would have a high frequency of generating epistemically virtuous states in our world."[3] In our view, this sort of objection trades on taking the analysis of a concept as tantamount, in paradigm cases anyway, to providing an objective decision procedure, a set of rules on the basis of which the acceptability of hypotheses can be assessed in the light of available data. But an analysis of concepts does not provide such a procedure. Nor is there any reason to demand that it should. The point of the analysis is, as just noted, clarification and explanation, not a prescription for doing good science. As for the "naturalist" claim that epistemology should be applied evolutionary biology or cognitive psychology if it is to have any value, it is enough for our purposes to point out, first, that this is itself a normative claim, resting on arguments whose validity is assumed to be independent of the outcome of biological or psychological investigations, and second, that the true test of a philosophical claim is whether it provides insight and has little to do with whether it originates in "arm-chair" reflection or is in some sense "a priori."[4] In footnote 11 in Chap. 1, we noted that Alvin Goldman, the progenitor of "reliabilism," still the epistemological theory most favored by naturalists, blithely assimilates "adequate evidence" with "highly warranted inductive inference" and thereby locates himself squarely, if also ironically, among traditional theorists.[5] We happen to think that science is more credible than most human cognitive activities, that it generates sufficiently good approximations to what we term "reality," and that some theories are better approximations than others. But our goal now is not to celebrate a particular methodology, or arbitrate between scientific hypotheses, or make a case for an ideal of objectivity or rationality, but to help resolve some long-standing epistemological difficulties.

Probabilistic Measures of Empirical Support Are Inevitably Dogmatic

The term "probability" has multiple definitions.[6] A number of critics have opined that the definition of probability as a measure of belief will inevitably prove to be inadequate. It has many variants. The eminent statistician Barnard (1949) argued that belief probabilities were unscientific when he said: "To speak of the probability of a hypothesis implies the possibility of an exhaustive enumeration of all possible hypotheses, which implies a certain rigidity foreign to the true scientific spirit.

---

[3]Kitcher (1992, pp. 75–76).

[4]"Naturalism" in epistemology to this point remains little more than a program, with few concrete results or basic insights other than those of Tversky and Kahneman mentioned in the preceding chapter. Hatfield (1990) exposes some of the difficulties when the program is actually pursued rigorously, as it was by Helmholtz in his theory of spatial perception.

[5]Indeed, he thinks that both traditional and naturalized perspectives have a place in epistemology.

[6]See *Our Two Accounts and Interpretations of Probability* in Chap. 2.

We should always admit the possibility that our experimental results may be best accounted for by a hypothesis which never entered our own heads." These sentiments are echoed in the philosophical literature by Philip Kitcher:

> …there is an interesting challenge to probabilistic notions of empirical support, since, when there are uncountably many rival hypotheses, it is impossible that all of them should be assigned non-zero probability. Does this make a certain dogmatism unavoidable?[7]

There are at least three different ways in which to mitigate if not also solve the problem. Each has to do with limiting the number of hypotheses or models under consideration.

A determined Bayesian could circumvent the problem by declaring that her probabilities are conditional on the true model being among the explicitly stated alternatives.[8] While technically correct, this way of mitigating it is less than satisfactory because probabilities would then no longer represent beliefs in hypotheses or models *tout court*, and the declaration would amount to begging the fundamental question.

A more "objective" Bayesian could rank order hypotheses in virtue of their respective simplicity[9] as measured in terms of the number of their parameters, for example, (or some other similar epistemic-pragmatic criterion), and then start by testing at the top of the order. Although no one has ever characterized "model-simplicity" in a generally acceptable way, the intuitive notion does have several important instrumental values for science.[10] On the one hand, simple models are more precisely estimable than complex models and, on the other hand, simple models are generally more understandable. These values enhance the two primary functions of models in science: to make predictions and to provide explanations.

Unfortunately, while the estimates of simple models are less variable than those of more complex models, there is a well-known trade-off between bias and precision with model complexity (Bozdogan 1987). In general, for every simple estimated model, a more complex model can be found which is less biased for prediction. Thus for the predictive use of models, at any rate, the benefit of estimation precision is offset by a cost due to reduction in prediction accuracy.

---

[7]Kitcher (2001, p. 37, n. 5). It illustrates the ambiguity of "probability" that what Barnard takes as the misleading character of the expression "the [measurable] probability of a hypothesis" is itself the flip side of the claim that no hypothesis is ever more than merely "probable" [i.e., might be mistaken].

[8]During the late 19th century, there were two mutually exclusive and jointly exhaustive theories of the nature of light, the corpuscular and wave theories. Now we know that this way of dividing the alternatives was mistaken, and that light has a wave-particle duality. The more we know about light, the more finely we should be able to partition possible competing theories.

[9]Or some other epistemic-pragmatic criterion. See Rosenkrantz in Earman (1983) for an argument in behalf of the "evidential value" of simplicity. See Bandyopadhyay and Brittan (2001) for a survey of several criteria of model-selection and an argument for adopting one in particular. See also Forster (2000) for some novel proposals.

[10]What follows is taken from Lele and Taper (2012). See also Fishman and Boudry (2013).

Similarly, for the explanatory use of models, the benefit of comprehensibility that comes with model simplicity is offset by the cost of lack of comprehensiveness. We return to this "curve-fitting problem" in the final paragraphs of the monograph. For the moment, it is enough to say that an appeal to "simplicity" alone is a slender, although not necessarily a broken, reed on which a Bayesian might limit the number of potential hypotheses to be at least in-principle under consideration.

Finally, Kitcher himself dismisses the problem, and with it all similar attempts to undermine objectivity on the basis of the fact that no matter how extensive, the data will always be compatible with an uncountable number of hypotheses, by appealing to the practice of those engaged in assessing hypotheses. "They tacitly suppose that there's a class of relevant alternatives and that fault lies in dismissing one of *these* without adequate evidence" (Kitcher 2001, pp. 37–38). There is no doubt that some hypotheses are intuitively more "relevant" than others, and that therefore they should be taken seriously. Nor is there any doubt that scientists focus their attention on only a small number of hypotheses at any given time. The problem, as Barnard indicates, is that a hypothesis initially neglected as "irrelevant" may prove to be the one which best accounts for the experimental results.

In our view this difficulty for purely probabilistic accounts of empirical support has not yet been dispelled (if in fact it is dispellable). We would only point out that the evidential approach to support avoids it completely by keeping its inferences local, that is, between alternative models considered pair-wise. This is a chief reason why we think that a Bayesian account of confirmation alone is not sufficient. The way in which data constitute evidence for and against hypotheses must also be taken into consideration. The ways in which taking it into consideration solves at least some of the classical dilemmas concerning the under-determination of hypotheses by data will be taken up again and at greater length in the section on the Grue Paradox in Chap. 9 and in all of Chap. 10.

On the Bayesian Account, Everyone Is (Wrongly) Justified in Believing All Tautologies

The third objection has to do more specifically with our account of confirmation. Our concept of evidence is not in terms of raising or lowering the belief probabilities of hypotheses on the basis of the data gathered, and to this extent our position is not simply "probabilistic," still less "Bayesian," but our concept of confirmation is. At first glance it seems little more than refined common sense that the more data we have for a claim the more justified we are in believing that it is true, but it has often encountered the objection that it leads to unintuitive results. There are two subsidiary lines of argument. One has to do with some of the beliefs allegedly "confirmed," the other with the improbable "grounds" of such confirmation. We will take up the first line of argument here, the second in the next section on the fourth objection.

Probabilistic analyses of confirmation (or here "justification") are often thought to incorporate "the simple rule:"[11]

A person is justified in believing P if and only if the belief probability of P is sufficiently high.

A direct defense of the simple rule has been made by any number of distinguished epistemologists,[12] many of whom are not Bayesians of any stripe. But it has also been very much criticized when it is framed in probabilistic terms.

> A reasonably familiar objection is that it follows from the probability calculus that every tautology has probability 1. It would then follow from the simple rule that we are justified in believing every tautology. Such a conclusion is clearly wrong. If we consider some even moderately complicated tautology such as $[P \leftrightarrow (R \vee -P)] \rightarrow R$, it seems clear that *until we realize that it is a tautology*, we are not automatically justified in believing it. The *only* way to avoid this kind of counterexample to the simple rule is to reject the probability calculus, but that is a very fundamental feature of our concept of probability and rejecting it would largely emasculate probability [understood in a Bayesian way, in terms of degrees of belief].[13]

This line of criticism rests on three assumptions, all of which are at the very least controversial: that the notion of "believing a tautology" makes sense, that tautologies (and all so-called "logical truths") are *true*, and that we are justified in believing any proposition only if we are aware of ("realize") the way in which we come to believe it. These assumptions can be rejected. That is, rejection of the probabilistic calculus (as measuring degrees of belief) is not the "only" way to avoid the counterexample indicated. Our task now is to show what is wrong with the first two.[14]

We take it that "believing a tautology" is tantamount to "believing that a tautology is true." This assumes, first, that "belief" is appropriate to with respect to tautologies (or, for that matter, the whole class of logical truths). It is not. The concept of belief is parasitic upon the concept of truth.[15] By this we mean that at least part of what is involved in attributing beliefs to oneself or others is to mark out a distinction between what is believed (to one degree or other) and what is true. That is, what is crucial to the concept of belief is the ever-present possibility that

---

[11]See Pollock and Cruz (1999, pp. 101ff).

[12]See Chisholm (1957, p. 28) and especially Kyburg (1974), *passim*, among many other classic sources.

[13]Pollock and Cruz (1999, p. 105), second emphasis ours.

[14]The third assumption is often known as "internalism," that a person must be in possession of the grounds on which a belief is held if she is justified in believing it. It is captured in our Bayesian account of confirmation. But our account of evidence is "externalist," i.e., whether or not data constitute evidence is independent of what an agent knows or believes.

[15]The line of argument here has been indicated most clearly by Donald Davidson. See, for example, his (1982). Very possibly Wittgenstein had something like this line of argument in mind at *Tractatus* 5.1362 when he writes "('A knows that p is the case', has no sense if p is a tautology)." Pears and McGuiness translation.

one might be wrong. One cannot have a "belief" unless it is possible that it is *false*. It follows at once that one cannot *believe* truths of logic (or their substitution instances), nor, a fortiori, tautologies. For truths of logic, if true, cannot possibly be false.

"Believing that a tautology is true" assumes, second, that truths of logic are *true*. It is, of course, a presupposition of the usual classical semantics for first-order languages that "truths of logic" *are* true. But this presupposition can be questioned. If the apparent "truth of logic" has the form of an identity, $a = a$, for example, and the singular term a does not refer, then there are good arguments to the conclusion that the identity is either truth-valueless or false, and so on for other atomic contexts in which singular terms do not refer.[16] It does not follow in one step from our realization that a given proposition is a tautology that we believe that it is true, for we might very well require, *inter alia*, that whatever singular terms it contains must refer before we assign it a truth-value.

The simple rule in its probabilistic guise, taking probabilities as degrees of belief, does not need to be rejected for the reasons given.[17]

Probabilistic Accounts of Justification (Wrongly) Hold that All Data Are Certain

The fourth objection is, like the third, against a generally probabilistic account of confirmation, and is perhaps a main reason why relatively few mainstream contemporary epistemologists, even philosophers of science, take belief-probabilities as seriously as they should. Pollock and Cruz (1999, p. 102) formulates it as follows: On Bayes Theorem,

> When we come to acquire new data Q, it (*sic*) will come to have probability 1. This is because prob $(Q \mid Q) = 1$. But what is it to acquire new data through, for instance, perception?...The beliefs we acquire through perception are ordinary beliefs about physical objects, and it seems most unreasonable to regard them as having probability 1. Furthermore, it follows from the probability calculus that if prob $(Q) = 1$, then for any proposition R, prob $(Q \mid R) = 1$. Thus if perceptual beliefs are given probability 1, the acquisition of further data can never lower that probability. But this is totally unreasonable. We can discover later that some of our perceptual beliefs are wrong.

When we have uncertain data, subject to correction in the light of subsequent experience, it makes little sense to assume, as applications of Bayes Theorem apparently must, that the data have probability =1, which is to say that

---

[16]This is to take a "free-logical" way with logical truth. See Meyer and Lambert (1968).

[17]Although we do so for other reasons. See the section on *Absolute and Incremental Confirmation* in Chap. 2.

they are certain. Sometimes the point is put in the form of a paradox: probabilistic approaches to confirmation claim that no empirical propositions are certain, yet must assume that some empirical propositions are certain if conditionalization on them is to be possible.

The second half of the paradox is freely acknowledged by at least some epistemologists. Thus C.I. Lewis in a famous passage: "If anything is to be probable, then something must be certain. The data which themselves support a genuine probability, must themselves be certainties" Lewis (1946, p. 186). Lewis avoids paradox by denying its first half, that is, he asserts that at least some empirical propositions, those having to do with what is immediately given in sense experience as against ordinary beliefs about physical objects *are* certain, and thereby provide suitable foundations for knowledge.

So far as our argument in this monograph is concerned, it does not much matter whether the data invoked in applications of Bayes Theorem are eventually subject to revision or not. It is enough for almost all of our purposes to point out that inference is hypothetical in character. If the premises are true, and the inference is deductively valid, then the conclusion is true. Whether the premises are true or not is a separate matter. In the same way, if the data are as given, and the posterior probability of a hypothesis is thereby raised, then to this extent the data justify or confirm the hypothesis. Confirmation in our view has to do with the relation between data and hypothesis, and not with the character of the data considered in and of themselves, even though the account of conditionalization so far roughly sketched presupposes that the data are to be taken at face value.

Most classical accounts of hypothesis testing presuppose the very same idea, that the data from which conclusions are drawn are not themselves subject to question. Only the inferential structure of such testing is at issue. For example, on the hypothetico-deductive model of confirmation observational consequences are derived from a hypothesis (or, more usually, hypotheses) either directly or together with so-called "correspondence rules," "bridge laws," or "operational definitions" linking theoretical concepts in the hypotheses with concepts more directly descriptive of experience in the observation statements. If the derived consequence is in fact observed, then to that extent the hypothesis is confirmed, if not, then the hypothesis is disconfirmed. No attempt is made, within the model, to assess the quality of the data. It is similarly the case with such other methods of theory testing as error-statistics. This is not to defend such methods, still less to argue that every genuine epistemic question can be resolved using them. It is to underline the fact that our use of conditionalization in our analysis of justification does not commit us uniquely to assuming that the data are correct, and not subject to further analysis and revision.

But Lewis was undoubtedly trying to make a more general point. It was not simply that in all inferences, the premises must be assumed, but that knowledge is possible only if what is "assumed" is also self-evident or certain in the sense that it cannot be denied (without contradiction) or doubted. This is to say that Lewis was a foundationalist. Claims about physical objects, which are never more than merely probable, can be justified only if the data supporting them are not, even in principle, subject to revision. We do not share this view. Hypotheses can be justified, indeed knowledge is possible, even if data-premises are themselves uncertain. This claim has two dimensions.

First, our way with data incorporates what Skyrms (1986) calls the "certainty model;" for the sake of simplicity, the data on which we conditionalize are taken to have Pr = 1. But as he goes on to point out, there are other "fallibility" models of conditionalization available, in fact an infinite number of them, among which so-called Jeffrey Conditionalization (Jeffrey 1990, Chap. 11), on which probabilities are not assigned to data themselves but are re-assigned to hypotheses as the result of some (apparently non-propositional and therefore neither true nor false, still less "certain" or "uncertain") experiential inputs. Fallibility models in general, Jeffrey Conditionalization in particular, offer ways out of Pollock and Cruz's argument. But we are wary of the idea that experiences per se can or should lead to a re-distribution of probabilities over hypotheses. From our point of view, what justifies us in believing something are the *facts* of the case, states of affairs described in particular ways. That the planets trace elliptical orbits around the sun justifies us in believing that Newton's Law of Universal Gravitation is true, and of course the fact of elliptical orbits rests on the observations that both Brahe and Kepler made. As we acquire more facts, we re-distribute probabilities over hypotheses accordingly. It is entirely consistent with our view that some of what we take to be "facts" at any given time turn out to be false. Nothing is really certain, knowledge has no bed-rock foundations. But our probability distributions are on the basis of the data in hand, the facts as we have them, to be revised as more and better information comes along.[18]

Second, there are many techniques available to bring data under some control, even if there is no way in which to guarantee their certainty. In the sciences we generally take as exemplary of objectivity, these data consist of measurements. The difficulty is that such measurements are invariably subject to error.[19] Two simple examples illustrate the point. First, it is an axiom of the measurement of length that if some object a is the same length as b and b is the same length as c, then a is the same length as c. We could scarcely proceed to measure objects in a coherent way if we did not assume that sameness of length was transitive. The difficulty is that in

---

[18]In this we follow Howson and Urbach (2006, p. 287): "In our [Bayesian] account there is nothing that demands what [are] taken as data in one inductive inference cannot be regarded as problematic in a later one. Assigning probability 1 to some data on one occasion does not mean that on all subsequent occasions it needs to be assigned probability 1." See also Levi (1967, p. 209). We use conditionalization as an eminently clear way of making a distinction between confirmation and evidence, not to defend it as a principle of rationality. For objections to it, and to the certainty and fallibility models, see Bacchus, et al. (1990). It should be noted that in fact more and more scientific testing does take data uncertainty into account, and our account is easily accommodated to it. More important conceptually is the possibility of misleading evidence, discussed in Chap. 8.

[19]What follows is indebted to Kyburg's work, in particular to Kyburg (1984). It remains a scandal that so many philosophers of science ignore both the importance of error and the sophisticated statistical techniques that have been developed to deal with it.

repeated trials, observers report that although *a* looks to be the same length as *b*, and *b* the same length as *c*, *a* and *c* do not appear to be the same length.

Third, and here we appeal to no more than one's elementary carpentry experiences, only very rarely do one's measurements of a particular length agree if the ruler allows tolerances of 1/32, even 1/16, of an inch. Errors happen. The problem is not to avoid them entirely, which is impossible no matter how many pains one takes in setting up experiments, but to quantify their appearance and then to develop statistical tools that will allow us to systematize the data in such a way that they can be used to compare and in confirmation of individual hypotheses. Henry Kyburg suggests two such statistical tools—the *minimum rejection principle*, which allows us to take the frequency of observational error to be "the least we *must* assume" in order to reconcile our observations with such axioms as transitivity of length, and the *distribution principle*, which "directs us to suppose that the distribution of errors among the various categories of statements is as uniform as possible" (given satisfaction of the minimum rejection principle).[20] Such principles do not provide for the detection of individual errors, but they do give us the relative frequencies with which they occur; once we have the *mean* of a series of measurements in hand, the standard deviations from it, and certain other statistical facts,[21] we can proceed to make inferences from the data. The point is that, using the statistical principles just outlined and/or many other others, one can tame if not also avoid error, to the point where data can be used to up-date probabilities and provide evidence even in the face of measurement error.

Of course, questions of judgment enter in.[22] If a well-trained physicist makes careful measurements that suggest some hypothesis might be false,[23] then we have to begin to take that possibility seriously. If a first-course physics student makes measurements that are outside the predicted values, we conclude that she needs to spend more time in the lab perfecting her skills. As in all science, there is the need to balance the demands of theory, for example, the theory of measurements of length (on which

---

[20]*Ibid.*, p. 91.

[21]As Kyburg notes, one does not have to be a statistician to know that "large errors are much less frequent than small errors; that errors tend to average out; and that [in day to day carpentry, say] an error of a quarter of an inch is relatively infrequent." *Ibid.*, p. 257. In the case of at least one of the authors, if in a series of measurements of a board to be cut we get the same value for two consecutive measurements, we call it good and proceed to saw away.

[22]Patrick Suppes (1984, p. 215), like Davidson and Kyburg one of our mentors, issues a caution that must be taken seriously with respect to our own and others' examples. "Published articles about experiments may make it seem as if the scientist can in a simple, rational way relate data to theory in the finished form commonly found in the scientific literature. In my own experience, nothing can be further from the truth. In the setting up of experiments and in the analysis of the results there are countless informal decisions of a practical and intuitive kind that must be made".

[23]I.e., measurements that are outside the predicted interval of values for the quantity in question, not measurements that are incompatible with the absolute value in the hypothesis, for only occasionally, and in a statistically predictable way, will measurements coincide with the absolute value.

transitivity, additivity, and so on, hold) and the results of experiment and observation, and no way to say in principle how in particular cases this balance is to be determined.

Just as the simple rule does not have to be rejected when we construe confirmation in terms of up-dating probabilities, so too we do not have to abandon the means by which we carry out such up-dating if, as is inevitably the case, the data with which we do so are not error-free or in the relevant sense "certain."

The Way Forward

Thus ends Part I, the clarification, illustration, and defense of our distinction between evidence and confirmation. In Part II, we examine four notable other ways in which evidence and confirmation have been understood, the first a modification of the classical Bayesian tradition, the other three very critical of that tradition.

References

Bacchus, F., Kyburg, H., & Thalos, M. (1990). Against conditionalization. *Synthèse, 85,* 475–506.

Bandyopadhyay, P., & Brittan, G. (2001). Logical consequence and beyond: a look at model selection in statistics. In J. Woods & B. Hepburn (Eds.), *Logical consequence.* Oxford: Hermes Science.

Barnard, G. (1949). Statistical inference. *Journal of the Royal Statistical Society: Series B, 11,* 115–149.

Bozdogan, H. (1987). Model selection and akaike information criterion (AIC)—the general theory and its analytical extensions. *Psychometrika, 52,* 345–376.

Chisholm, R. (1957). *Perceiving.* Ithaca, NY: Cornell University Press.

Davidson, D. (1982). Rational animals. *Dialectica, 36*(4), 317–327.

Earman, J. (Ed.). (1983). *Testing scientific theories, Minnesota studies in the philosophy of science* (Vol. X). Minneapolis: University of Minnesota Press.

Fishman, Y., & Boudry, M. (2013). Does science presuppose naturalism (or anything at all)? *Science & Education, 22,* 921–947.

Forster, M. (2000). The new science of simplicity. In H. Kreuzenkamp, et al. (Eds.), *Simplicity, inference, and economic modeling.* Cambridge: Cambridge University Press.

Hatfield, G. (1990). *The natural and the normative.* Cambridge, MA: MIT Press.

Howson, C., & Urbach, P. (2006). *Scientific reasoning: the bayesian approach* (3rd ed.). Chicago and LaSalle, IL: Open Court Publishing.

Jeffrey, R. (1990). *The logic of decision.* Chicago: University of Chicago Press.

Kitcher, P. (1992). The naturalists return. *The Philosophical Review, 101*(1), 53–114.

Kitcher, P. (2001). *Science, truth, and democracy.* Oxford: Oxford University Press.

Kuhn, T. (1962/1970/1996). *The structure of scientific revolutions.* Chicago: University of Chicago Press.

Kyburg, H. (1974). *The foundations of scientific inference.* Dordrecht: D. Reidel.

Kyburg, H. (1984). *Theory and measurement.* Cambridge: Cambridge University Press.

Lele, S., & Taper, M. (2012). Information criteria in ecology. In Hastings & Gross (Eds.), *Sourcebook in theoretical ecology.* Berkeley: University of California Press.

Levi, I. (1967). Probability kinematics. *British Journal for the Philosophy of Science, 18,* 200–205.

Lewis, C.I. (1946). *An analysis of knowledge and valuation.* LaSalle, IL: Open Court.

Meyer, R., & Lambert, K. (1968). Universally free logic and standard quantification theory. *Journal of Symbolic Logic, 33,* 8–26.

Pollock, J., & Cruz, J. (1999). *Contemporary theories of knowledge* (2nd ed.). Totiwa, NJ: Rowman and Littlefield.

Rosenkrantz, R. (1983). Why Glymour is a Bayesian. In Earman.

Skyrms, B. (1986). *Choice and chance*, (3rd ed.). Belmont, CA: Dickenson.

Suppes, P. (1984). *Probabilistic metaphysics*. Oxford: Blackwell.

Part II
Comparisons with Other Philosophical Accounts of Evidence

Chapter 5
A Subjective Bayesian Surrogate for Evidence

Abstract We contend that Bayesian accounts of evidence are inadequate, and that in this sense a complete theory of hypothesis testing must go beyond belief adjustment. Some prominent Bayesians disagree. To make our case, we will discuss and then provide reasons for rejecting the accounts of David Christensen, James Joyce, and Alan Hàjek. The main theme and final conclusions are straightforward: first, that no purely subjective account of evidence, in terms of belief alone, is adequate and second, that evidence is a comparative notion, applicable only when two hypotheses are confronted with the same data, as has been suggested in the literature on "crucial experiments" from Francis Bacon on.

Keywords Subjective bayesianism · S measure of confirmation · Accumulation of evidence · Paradox of subjective evidence

Christensen's S Measure of Confirmation

Christensen is interested both in developing an account of how data support (confirm, provide evidence for) a hypothesis and in determining which probabilistic measure appropriately represents the support relation between data and hypothesis.[1] One way to understand his account is to contrast it with a standard measure, derivable in probability theory that captures the effect of the improbability of a datum on raising confidence in a hypothesis, H. The lower the marginal probability of the datum [$\Pr(D) \neq 0$], the greater the effect on raising an agent's confidence in

---

[1]Christensen has shown that another measure, $S^*(D/H) = \Pr(H/D) - \Pr(H)$, is equivalent to $S(D, H)$. Joyce arrived at the same measure independently, and Hájek and Joyce (2008) later called it "probative evidence," denoting it by $\mathbf{q}(,)$.

H, i.e., in raising the posterior probability of *H*. Christensen thinks that the potential increase in Pr(*H* | *D*) is a consequence of two factors: the degree of "evidential tie" *D* has to *H* and the distance Pr(*D*) has to travel to attain probability 1.[2] Since he is interested in the "evidential tie" between *D* and *H,* Christensen wants to control the effect of the second factor, viz., how far *D* has to travel in attaining probability 1, viz., [1– Pr(*D*) = Pr(⋅*D*)]. After normalizing the measure by dividing by Pr(~ *D*), the measure of (confirmational) support he arrives at is:

$$S(D,H) = \frac{\Pr(H|D) - \Pr(H)}{\Pr(\sim D)}$$

where $0 < \Pr(D) < 1$.

Like many Bayesians, Christensen holds that Pr(*D*) never equals 1 since *D* is an empirical proposition and empirical propositions are always possibly false. In contrast, Joyce leaves room for cases in which Pr(*D*) could increase or decrease continuously to 1 or 0. His allowing for the increase/decrease of Pr(*D*) to 1/0 is related to the motivation of his evidence relevance measure **q** (,). He couches his measure in terms of a comparison of "confirmational power" between two propositions, that of *D* and its negation. Confirmational power involves comparing an agent's degree of belief in a hypothesis *H* upon learning that *D* is true to her degree of belief in *H* on learning that ~ *D*. The comparative difference in the agent's degrees of belief is then reflected in his **q** (,), which equals Pr(*H* | *D*)– Pr(*H* | ~*D*).[3] For the present discussion, however, what matters are the similarities between Christensen and Joyce's approaches to confirmation issue. The most important of

---

[2]Christensen's own candor needs to be acknowledged at the outset. What he has provided, he says (Christensen 1999, p. 460), is not an "account of confirmation", but a way of understanding certain features of it. Shedding even a little light is better than shedding no light at all.

[3]Joyce (1999) handles the case where Pr(*D*) reaches its highest value by considering a Reyni-Popper measure, on which (as against the standard Kolmogorov definition), Pr(*H* | ~*D*) is defined when Pr(*D*) = 1. Joyce finds it quite counter-intuitive that when the agent's value for *D* changes continuously and ultimately attains value 1, due to her learning new information, the confirmational value of *D* for her will stop suddenly. The intuitive way to approach the case where Pr(*D*) reaches 1, according to him, is to consider a specific sense in which *D* could still be counted as evidence for *H* even when the agent is certain about *D*. On his interpretation, *D* is evidence for *H* in that *H* is more likely to be the case given *D* (whatever its probability) than not-*D*. He thinks that this sort of intuition underlies the confirming power of "old evidence" (see Chap. 9 for a discussion of the "old evidence" paradox). Fitelson (1999) notes, however, that if one incorporates a Reyni-Popper measure in **q**(,) or its equivalents, then **q**(,) differs not only from the conventional quantitative Bayesian measure of confirmation (as they do not on the Kolmogorov definition), but also from the purely qualitative measure. Indeed, Fitelson argues that if one incorporates the Reyni-Popper measure, there are many possible qualitative "Bayesian" measures of confirmation, and just as many "old evidence problems" (only one of which the Christensen-Joyce measure addresses). Since a principal motive of the C-J account is to resolve "the" old evidence problem, this is a serious difficulty for it. In trying to close one door, many others appear to have opened.

these similarities is that for both of them confirmation and evidence can be understood uniformly in a Bayesian way as subjective probability measures, and thus that they can be folded into the same conception.

Intuitions and Purely Probabilistic Measures

Christensen is careful to note that his S-based measure of evidence fails to match pre-theoretic intuitions. Furthermore, he thinks that this shortcoming exists in general for any formal probabilistic account of confirmation/evidence. He writes, "The reasons for the mismatches suggest that no purely probabilistic account could match those (intuitive) judgments" (Christensen 1999, p. 460). He construes "purely probabilistic" in a wide sense to include even measures that are characterized in terms of probabilistic concepts but whose values are not restricted to the 0–1 range. Thus he claims, for instance, that the ratio of likelihoods measure, the basis of our account of evidence, fails just as much as the S-measure does to match our intuitive confirmational/evidential judgments. We will briefly discuss two reasons he advances for this claim, and argue that they do not apply to our account of evidence.

His first reason is that the common-sense concept of evidence and support is "indeterminate". Christensen feels that an account of evidence that is limited to a single evidence/support function misses something important about our intuitive judgments about the concept, namely, that we habitually ask a number of different questions, not just one, when we ask how data provide evidence for or support a hypothesis. They are in this sense "indeterminate" notions. We agree. The recent literature on evidence (e.g., Lele 2004; Claeskens and Hjort 2008; Taper et al. 2008) has stressed that a wide variety of evidence functions are both possible and necessary in the face of a wide variety of evidential questions. Consequently, the formal evidential account, of the kind proposed in the section "The Evidential Condition" of Chap. 2 is able to handle this wide variety of evidence-questions. We will discuss this point by adapting an example from Christensen for our purpose.

In his example, Christensen notes that the question "[how much] is a candidate C supported by an interest group I?" is indeterminate because a variety of more specific questions could be intended, for instance, "[what is] the number of dollars C received from I?" or "what proportion of C's funds is supplied by I?" Christensen's insight is that scientists and philosophers are more interested in certain features of the data or underlying process than they are in the data or process as a whole. Fortunately, evidence functions can be constructed using any transformation of the data or any functional[4] of the data distribution. Thus, not only can one ask which model best matches the data as a whole, one can also ask which

---

[4]A functional is a "function" of an entire distribution such as the mean, variance, or proportion in a category.

model recapitulates certain specified features of the data such as the total amount or proportion of donations greater than a threshold as in Christensen's questions.[5]

It is perfectly reasonable to apply different evidence functions to the same data and set of models, and perfectly plausible to find different models with the greatest degree of support if we do. For instance, researchers may want to ask which model is most similar to the underlying process that generated the data, or they may wish to ask which model will best predict a new data point.[6] The same model may not provide answers to these two questions.[7]

So, contrary to Christensen's general claim that the common-sense notion of support is indeterminate and that any formal (and therefore "determinate") account won't be able to handle a variety of cases misses the mark in two respects: first, different evidence functions can be chosen based on the specific evidence question being asked, and, second, all of these evidence functions can be distinguished from a Bayesian confirmation function.

Consider Christensen's second reason for claiming that "purely probabilistic" accounts do not match our intuitive judgments concerning the support data provide hypotheses. The relationship represented by S or any other formal account of evidence is symmetric as, according to him, the data confirm a hypothesis just in case the hypothesis confirms the data.

$$[\Pr(H|D) - \Pr(H)] > 0 \leftrightarrow [\Pr(D|H) - \Pr(D)] > 0.$$

This is possible because in such frameworks, including our own account of confirmation, both data and parameters are implicitly treated as random variables (i.e., described by probabilities with specified priors). Christensen continues that on our common-sense notion of evidence and support, the evidential relationship is asymmetrical. For example, the bending of light in the vicinity of a massive object like the sun provides evidential support for the General Theory of Relativity (GTR), but the GTR does not provide evidence for the bending of light. In the evidential framework we advocate, observed data, although generated by a stochastic process, are assumed known and thus fixed,[8] while models are assumed to be fixed data-generating mechanisms which have an unknown relationship to the "true" data-generating process. Thus data can support one hypothesis over another, but they need no support because they are (on our assumption) already known. Because they are known, and hypotheses, although fixed, are at an unknown distance from the truth (whatever that might prove to be), the relationship between data and competing hypotheses is intrinsically asymmetrical in our evidential framework as well.

---

[5]See particularly the discussion of the Focused Information Criterion (FIC) in Claeskens and Hjort (2008).

[6]Even though prediction is involved, this is an evidential question as soon as the researcher asks the comparative questions, "which model has the greatest predictive power?"

[7]See Bozdogan (1987) for a thorough statistical discussion of the point and Taper and Gogan (2002) for an ecological example.

[8]See our earlier discussion of this assumption in Chap. 2.

Neither of the worries that Christensen thinks plague formal "purely probabilistic" accounts of confirmation and evidence tell against our account of the latter. As we have argued, our account incorporates certain deep intuitions about the notion of evidence, and is at the same time able to handle a variety of questions about support with the help of different types of evidence functions.

The Subjective Bayesian Account and Ours Compared

Now to make clear the main differences between what we are calling the "subjective Bayesian account" of confirmation and evidence and our own.[9] According to their account, when confidence is high or low, additional evidence does not necessarily affect any significant change in an agent's degree of belief in a hypothesis ("confidence" used interchangeably with "degree of belief"). Borrowing Christensen's example,[10] suppose an agent has formed a hypothesis that there is a deer nearby, and has observed fresh deer droppings. The droppings provide high confirmation, other things being equal, for the deer-presence hypothesis. As we know, an agent's up-dated degree of belief becomes her *new* prior probability for the deer hypothesis, which is now very high. On subsequently observing an antler nearby, the agent's confidence in the deer hypothesis does not change significantly even though observing the antler (at the right time of year) is also strong evidence for the deer hypothesis. In the same way, when an agent's degree of confidence in a hypothesis is very low, strong evidence may not be able to raise it substantially.

A real-world example in physics of this resistance to changing confidently-held beliefs is the 2011 experimental claim that neutrinos travel faster than the velocity of light in a vacuum. In the OPERA experiments conducted in Italy, neutrinos (which have small mass), appeared to be traveling faster than the speed of light.[11] Before these results were called into question, many particle physicists, including both theoreticians and experimentalists, took them as ground-breaking *if they could be verified*, but did not have any confidence in the experiments originally conducted. Even before a glitch in the experiments was uncovered, the results were considered highly improbable; a speed higher than that of light in a vacuum would be a clear violation of a corner-stone of modern physics. Because the degree of confidence of a member of the particle-physics community in the hypothesis that neutrinos could travel faster than the speed of light was already so low, the apparently (though temporary) strong evidence for the hypothesis did not significantly affect the community's confidence that it was false. In such cases, the S-measure seems to encapsulate nicely the popular idea that extraordinary claims require extraordinary evidence.

---

[9]Our reconstruction is based mainly on Christensen (1999).

[10]Although not necessarily endorsing the "woodscraft" expressed.

[11]For a brief summary, see Brumfiel (2012).

Fig. 5.1 A plot of S given a
positive test result as a
function of the prior
probability that the patient has
the disease. We plot under 3
combinations of sensitivity
and specificity
(sensitivity = 0.9 and
specificity = 0.9,
sensitivity = 0.9 and
specificity = 0.7, and
sensitivity = 0.7 and
specificity = 0.9)

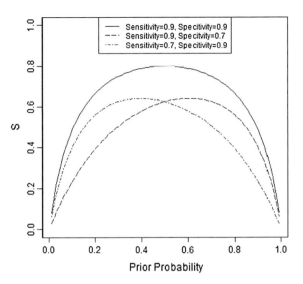

What those two examples imply is that S measure is sensitive to prior probabilities in a way that our conception of evidence is not. We can demonstrate the
sensitivity of S to prior probabilities in a schematic way by plotting S in screening
tests for diagnostic studies. The purpose of these screening tests (e.g., PAP smear)
is to determine the likelihood of having cervical cancer, so that physicians can
reduce morbidity by detecting diseases in their earliest stages when treatment is
usually more successful. The quality of a diagnostic test, as well as the values of S
and LR depend on the probabilities of possible test outcomes if the patient has the
disease and if she does not. Often these probabilities are quantified as the sensitivity
and specificity of the test. Sensitivity refers to a test's ability to designate an
individual with disease as positive. The specificity of a test is its ability to designate
an individual who does not have a disease as negative. It is desirable to have a test
that is both highly sensitive and highly specific. But frequently this is not possible.

In Fig. 5.1, we plot S given a positive test result as a function of the prior
probability that the patient has the disease. We plot under 3 combinations of sensitivity and specificity. In all cases S depends strongly on prior probability. S is
small if the prior probability is either high or low. The likelihood ratios for these
three cases are 9, 3, and 7 respectively, and are independent of prior probability.

Although it has been lurking at the edges of our discussion, we have not yet
made explicit the tension in Christensen's analysis between taking it as an account
of confirmation or of evidence. In the deer example, he correctly points out that
both observing deer droppings and, on a separate occasion, observing an antler are
evidence for the hypothesis that there is (or recently was) a deer nearby. Likewise,
from the perspective of our own account of evidence, observing deer droppings or
observing an antler counts as evidence for the deer hypothesis, and they can be
pooled to constitute still stronger evidence for the deer hypothesis than its
deer-absence alternative (For how our account of evidence is able to pool data, see

Chap. 8, section "Quantitative Thresholds, the Accumulation of Evidence and "Good Reasons to Believe"". His S-based account of confirmation, which exploits an agent's prior probability, is not, however, able to incorporate this evidential intuition. Once deer droppings are observed, observing antlers does not increase an agent's degree of belief in the deer hypothesis. An examination of the S-measure makes this clear. The deer example and the OPERA experiment illustrate what the S-measure really measures: *how degree of belief should be revised by data in the context of current beliefs*. In downplaying the importance of *new evidence*, it is an essentially conservative approach to the methodology of science.

The Subjective Bayesian Account: Inadequate on Evidence

The S-measure makes explicit the two key features of Christensen, Joyce, and Hájek's (CJH) position: (i) when an agent's degree of confidence in a hypothesis is high, intuitively strong evidence is not able to influence her degree of belief in it very significantly; (ii) when an agent's degree of confidence in a hypothesis is low, intuitively strong evidence is unable to contribute in a significant way to increasing her degree of belief in it. Thus on the CJH position, degree of confidence (belief) is to be distinguished from strength of evidence. But, we have argued, in downplaying strength of evidence, the S-measure does not provide a complete account of hypothesis-support, still less of hypothesis-testing.

We used the TB example to show that, yes, it is possible that a rational agent have a low degree of belief that a person has the disease, even though the evidence that the disease is present rather than absent is strong. It does not follow that confirmation and evidence are numerically unconnected. D confirms H just in case D provides at least some evidence for H against a rival hypothesis, so long as the hypotheses in question are mutually exclusive and jointly exhaustive. That is, $\Pr(H \mid D) > \Pr(H)$ if and only if $[\Pr(D \mid D)/\Pr(D \mid \sim H) > 1]$, a result owing to our colleague Robert Boik. This point bears in an instructive way on the counter-intuitive character of the S-measure.

Explicitly, the S-measure is not a comparative account of evidence in that only one hypothesis appears in its formulation. Implicitly, however, it could be conceived of in terms of a comparison between a hypothesis, H, and its denial, $\sim H$, because with an exhaustive pair of hypotheses, any increase in the probability of H implies a decrease in the probability of $\sim H$. We will argue with the help of a schematic example that if one's hypotheses are not binary, S turns out to be less than satisfactory as an evidence measure since it would provide evidence for a hypothesis which is not the most confirmed by the data.

Consider H_1, H_2, and H_3 to be three mutually exclusive and jointly exhaustive hypotheses in a domain with priors of 0.33 assigned to each of them. We use uniform priors over these competing hypotheses so that priors will have the least influence on which hypothesis should be most confirmed. Data D are observed. Let's assume $\Pr(D|H_1) = 0.6$, $\Pr(D|\sim H_1) = 0.4$, $\Pr(D|\sim H_2) = 0.3$,

$Pr(D|H_3) = 0.2$, and $Pr(D| \sim H_3) = 0.8$. Based on those values, we calculate $S(H_1) = 0.14$, $S(H_2) = 0.26$, and $S(H_3) = -0.4$. Based on the S-measure, two hypotheses have positive values; therefore, according to the Bayesian account of confirmation, there is increased reason to believe both H_1 and H_2. However, because S is not comparative, we have "reason to believe" a hypothesis that is not most supported by the data. When we use the LR-based account of evidence, we discover that only H_2 is the more likely hypothesis because the likelihood ratios indicate that given the data H_2 is supported 1.17 times as much as H_1 and 3.5 times as much as H_3.

On our analysis, the counter-intuitive character of the CJH position just indicated stems from its failure to distinguish sharply enough between confirmation and evidence,[12] which is to say that whatever account of "evidence" it might offer already presupposes an agent's subjective degree of belief, in particular her prior probability. On a deeper analysis, the CJH position fails to acknowledge the difference between the "what should I believe?" and "what is the evidence?" questions.[13] Christensen concedes that his account does not necessarily match our intuitive understanding of evidence, but he does not consider that a problem. Neither does he think that it is exactly an account of confirmation. But then we have to ask, what is its descriptive or normative force?

The Paradox of Evidence on the Subjective Bayesian Account

Christensen writes that it is paradoxical that "[t]he more confident we become in our evidence, the less it can *be* evidence". But it is only paradoxical if one conflates evidence and confirmation. Data with a high marginal probability change belief in hypotheses little, and thus have little confirmatory power. The evidential

---

[12]See Christensen (1999): "It [i.e., probability theory] provides a basis for conditionalization principles regulating change of belief—a topic about which traditional logic had little to say; it offers a quantitative analysis of our notion of confirmation, or evidential support". See also Joyce (1999): "all Bayesians agree that the degree to which D counts as evidence for or against H for a given person is a matter of the extent to which learning D would increase or decrease her confidence" (we have replaced Joyce's "X" by "H" and "C" by "D" in the quote to square with our usage), or again, "Relative to q(q,)...the extent to which D counts as evidence in favor of H for a given person depends on that person's degree of belief for H" (same letter replacement). As already noted in the text, Joyce uses "confirmational power" and "evidential relevance" interchangeably.

[13]Joyce (2004) might well question whether there is any worth to this objection because he denies that there is any issue. He would like to say that the various measures represent distinct, but complimentary notions of evidential support. Our response to his attempt to bring confirmation and evidence under the encompassing rubric of "evidential support" is, first, that these notions are intuitively distinct, second that the distinction exposes the motive for the scientific need to distinguish hypotheses (recall the global warming example) and, third, that blurring the line between them is the root cause of important epistemological problems.

relationship, as we and other evidentialists understand it, is independent of prior belief; whether or not we become more confident of our data does not change the way in which they provide or fail to provide evidence. The dependence of confirmation on prior belief generates another problem for our intuitive understanding of what counts as evidence. The degree of Bayesian confirmation depends on the order in which the observations are made. This is not necessarily the case in an assessment of evidence as we have characterized it[14] as we will demonstrate in the next chapter.

Revisit our discussion of the deer example. In it, observing deer droppings increased an agent's degree of confidence in the deer hypothesis. However, as the story proceeds, antlers are subsequently observed in the same area. Yet their sightings do not increase the agent's confidence in the deer hypothesis significantly, even though intuitively the antlers provide additional strong evidence for the hypothesis. Christensen is puzzled by this, that while observing antlers *is* strong evidence, it does not much increase the agent's confidence in the hypothesis once the deer droppings have been sighted. Our diagnosis for his puzzlement is that he wants to treat both deer droppings and antlers as evidence on the same par, but cannot given his account of confirmation via the S-measure.

We hope that this discussion of the Christensen, Joyce, and Hájek Bayesian confirmation measure illuminates not simply its independent interest and the ways in which we part company with it, but as well why an independent, non-Bayesian account of evidence is a necessary supplement to any "purely probabilistic" account of confirmation. What the S-measure does well is to reveal how belief should be changed by data in the context of other current beliefs. A case in point is the deer example. But it has the untoward consequences that we might very well have reason to believe a hypothesis which is not, as against its competitors, the most likely and that new evidence for it, however strong, is discounted.

In the next three chapters we turn to non-Bayesian accounts of evidence. A careful consideration of these accounts further strengthens our own.

References

Bozdogan, H. (1987). Model selection and akaike information criterion (AIC)—the general theory and its analytical extensions. *Psychometrika, 52*, 345–370.

Brumfield, G. (2012). Neutrinos not faster than light. *Nature,* 16 (March 16). doi:10.1038/nature.2012.10249.

Christensen, D. (1999). Measuring confirmation. *Journal of Philosophy, 99*(9), 437–461.

Claeskens, G., & Hjort, N. (2008). *Model Selection and Model Averaging.* Cambridge: Cambridge University Press.

Fitelson, B. (1999). The plurality of Bayesian measures of confirmation and the problem of measure sensitivity. *Philosophy of Science, PSA, 66*(1), 362–378.

---

[14]Likelihoods and evidence can of course depend on observation order if there is correlated observation error or some of the hypotheses represent conditional data-generating mechanisms.

Hájek, A., & Joyce, J. (2008). Confirmation. In S. Psillos & M. Curd (Eds.), *The Routledge Companion to Philosophy of Science*. New York: Routledge.

Joyce, J. (1999). *The Foundations of Causal Decision Theory*. Cambridge: Cambridge University Press.

Joyce, J. (2004). Bayes theorem. In *Stanford Encyclopedia of Philosophy*. http://plato.stanford.edu/entries/bayes-theorem/.

Lele, S. (2004). Evidence function and the optimality of the law of likelihood. In (Taper and Lele, 2004).

Taper, M., & Gogan, P. (2002). The northern yellowstone Elk: density dependence and climate conditions. *Journal of Wildlife Management, 66*(1), 106–122.

Taper, M., et al. (2008). Model structure adequacy analysis: selecting models on the basis of their ability to answer scientific questions. *Synthèse, 163*, 357–370.

Chapter 6
Error-Statistics, Evidence, and Severity

Abstract Several non-Bayesian and non-Likelihood accounts of evidence have been worked out in interesting detail. One such account has been championed by the philosopher Deborah Mayo and the statistician Ari Spanos. Following Popper, it assumes from the outset that to test a hypothesis is to submit it to a *severe* test. Unlike Popper it relies on the notion of error frequencies central to Neyman-Pearson statistics. Unlike Popper as well, Mayo and Spanos think that global theories like Newtonian mechanics are tested in a piecemeal way, by submitting their component hypotheses to severe tests. We argue that the error-statistical notion of severity is not adequately "severe," that the emphasis on piecemeal testing procedures is misplaced, and that the Mayo-Spanos account of evidence is mistakenly committed to a "true model" assumption. In a technical Appendix we deflect Mayo's critique of the multiple-model character of our account of evidence.

Keywords Error-statistics · Severe tests · Error-probabilities · Piecemeal testing · Global theories · The "true-model" assumption · Multiple models

Three Main Features of Error-Statistical Accounts of Theory Testing

There are a variety of objections to Bayesianism. We have already argued at length that it cannot provide an adequate account of evidence, and hence cannot provide a fully general and satisfactory characterization of hypothesis testing or support. The usual reason given by its critics is that the use of prior probabilities in a variety of contexts where relative frequencies, application of the Indifference Principle, or what are sometimes styled "plausibility arguments" are not available to determine them, introduce an element of unwanted "subjectivity" into the testing process. Consequently, alternative approaches to hypothesis testing and support have been developed. For the most part, they eschew subjective probabilities as we do in the case of evidence, but collapse the distinction between confirmation and evidence, i.e., maintain that data which constitute evidence for at the same time confirm

© The Author(s) 2016
P.S. Bandyopadhyay et al., *Belief, Evidence, and Uncertainty*,
Philosophy of Science, DOI 10.1007/978-3-319-27772-1_6

hypotheses, and thus explicate a conception of evidence which is very different
from our own.

One such alternative has been worked out in interesting detail by the philosopher
Deborah Mayo and the statistician Ari Spanos, among others.[1] It is known as the
error-statistical approach. Following Popper,[2] it assumes from the outset that to test
a hypothesis is to submit it to a *severe test*. Unlike Popper, Mayo relies on the
notion of error frequencies central to Neyman-Pearson error statistics.[3] The basic
idea is that a test is severe just in case it would detect an error in the hypothesis, that
is, the chances of its doing so are very high if in fact an error were present (and
would very likely not detect one if the hypothesis were true). Error frequencies
provide adequate information to compare hypotheses. Sellke et al. (2001) show that
for simple hypotheses, the likelihood ratio, P-values, and α (the size of a
Neyman-Pearson test) are transformable into each other. Thus properly considered,
error frequencies can constitute evidence, and one should be able to make reliable
statistical inferences about hypotheses based solely on them. It is in this way that
the approach is thoroughly objective.

Mayo breaks with Popper in two other respects as well. First, while he focuses
on the severe testing of global theories which consist of a number of different
hypotheses, she proceeds piecemeal, gaining experimental knowledge by local
arguments from error. "[W]hen enough is learned from piecemeal studies," she
writes, "severe tests of higher-level theories are possible"[4] and again, "[b]y building
up severely affirmed effects and employing robustness arguments, a single type of
local inference—once corroborated, can yield the theory."[5] In this and in other
ways, Mayo likes to say that error statistics is much closer to the "nitty-gritty" of
actual scientific practice than are, for example, the rather schematic Bayesian and
Popperian approaches. Second, error-statisticians contend that the passage of one or
more severe tests provides us with a good reason to believe that a hypothesis which
passes is true, as against Popper who holds that passage of severe tests allows us to

---

[1]See especially Mayo (1996, 1997a, b, 2005), Spanos (1999), and Mayo and Spanos (2006).

[2]Popper (1959). In addition to the terminology, Mayo incorporates (as does the Bayesian paradigm) the Popperian theme that the more surprising the experimental/observational outcome, the
better will be its value as a test of a particular hypothesis. This is not to suggest, however, that
Mayo simply takes over Popper's conception of severity. Note the following: First, Popper is a
deductivist; he thinks the rules of deductive logic suffice for capturing the idea of falsification (the
key to theory-testing on his account), whereas the error-statistical, Bayesian, and evidentialist
accounts are all very much non-deductivist in spirit (probability theory is crucial to an explication
of severity). Second, Popper links his idea that hypotheses can never be confirmed, only "corroborated" (not falsified to this point); but corroboration is exclusively a measure of past per-
formance and provides no indication concerning the success of a hypothesis in the future, whereas
Mayo thinks, as we shall see, that a test passed severely provides good reasons to believe that it is
true, i.e., will continue to hold of our experience.

[3]Which itself borrows from the "learning from errors" theory of Ronald Fisher. See Fisher (1930).
See also Pearson (1955), and Neyman (1976).

[4]Mayo (1996), p. 190.

[5]Mayo and Spanos (2010), p. 83.

say no more than that the successful hypothesis is not false. We will call the error-statistical contention that genuine evidence for a hypothesis provides us with a good reason to believe the hypothesis is true the "true-model" assumption. It is common to all of the theories of evidence we discuss in this monograph with the exception of our own. It is also, as we shall try to show in this and the following chapters in Part II of the monograph, very problematic.

Now to examine all three features of the error-statistical position—the account of severe testing, the emphasis on piecemeal procedures, and the "true-model" assumption.

Severe Tests

Mayo begins by characterizing strong evidence in terms of severity. "Passing a test (with D) counts as a good test or good evidence for H just to the extent that H fits D and is a *severe test* of H."[6] She terms this "the severity requirement" (SR). It is "always attached to a particular hypothesis passed or a particular inference reached."[7] Severity is the property of a test of a particular hypothesis H implicitly contrasted with its negation ~H with respect to background information B, auxiliaries A, and data D.[8] H passes a severe test T with outcome D just in case

1. "*D agree with or 'fit' H* (for a suitable notion of fit) and
2. Test T would (with very high probability) have produced a result that fits H less well than D does, if H were false or incorrect."[9]

Mayo does not provide a further specification of "fit." Any specification will do so long as it entails that $\Pr(D; H) > \Pr(D; \text{not-}H)$, where $\Pr(D; H)$ is *not* to be read as $\Pr(D \mid H)$, and is not to be construed as a *likelihood*.[10] In this connection, she quotes approvingly the engineer Yakov Ben-Haim, whose restatement of 2. is "We are subjecting a proposition to a severe test if an erroneous inference concerning the

---

[6]Mayo (1996), p. 180. She eventually replaced "*e*" with "*x*" in her formulation, signaling a distinction between "data" and "evidence." We will use our "*D*" to make comparisons between our accounts syntactically more perspicuous.

[7]Mayo (1996), p. 184.

[8]Mayo is not explicit about auxiliaries and background information in formulating her definition of a test's severity. But it is a standard assumption in present-day philosophy of science that a hypothesis given a datum is unlikely to yield a reliable probabilistic relationship unless we specify its auxiliaries and background information. However, how local or global testing of a theory via an auxiliary can be done remains unclear in her error-statistical account. We owe this point to Malcolm Forster.

[9]Mayo (2005, p. 99).

[10]$\Pr(D; H)$ is the probability of the data under a fixed hypothesis, $\Pr(D \mid H)$ is the probability of the data under a realized value of a random hypothesis, i.e., on a Bayesian account. Neither is *in sensu stricto* a likelihood. $L(H; D)$ is only proportional to $\Pr(D; H)$. There is an unknown constant that is removed by taking ratios.

truth of the proposition can result only under extraordinary circumstances."[11] Note that this condition makes the same sort of appeal to a general consensus concerning what constitutes "high" probability or "extraordinary" circumstances that is involved when we say that probabilities or likelihood ratios are "very high/low" or "very strong/weak," and that it is similarly "contextual" in that tests as such are not severe, but only relative to the particular inference that is claimed to have passed the test. The main conceptual and formal differences, rather, are that she does not include prior probabilities, likelihoods, and likelihood ratios in her characterization of severity. According to her, it is neither appropriate to do so since error-probabilities are to be distinguished from likelihoods, nor helpful since likelihood ratios cannot handle composite hypotheses,[12] nor necessary since the various objections to her account, which focus in part on her neglect of prior probabilities, can all be dissolved.

Reduced to its bare bones, here is the sort of intuitive case she has in mind. We test for the presence of a disease, say cervical cancer using the PAP smear. Suppose that when the disease is present, the test has a positive result over 80 % of the time; when the disease is not present, the test has a negative result over 90 % of the time. Given the high percentage of true positives if the disease is present, and the very low percentage of false positives if it isn't, the test would seem to be severe; it rules out that the subject has cervical cancer when she does not, and rules in, so to speak, that she does when she has it. Along the same lines, if the test were to result in the same percentage of false positives when disease is absent as true positives when disease is present, it would not be severe; a positive result would not rule out, to a high degree of probability, that the subject did not have the disease.

The Error-Statistical Account of Severity Tested

The basic idea that the severity of experimental tests can be used both to characterize and measure evidence is initially promising. For one thing, the search for severe tests, as Popper urged very effectively, does characterize scientific practice, and it is intuitive to connect this search with the strength of the evidence for hypotheses that it produces. For another thing, severity is "statistically correct;" a number of excellent statisticians[13] have vetted the mathematics and seen no problems. The issues concern the epistemological assumptions made and the implications drawn when evidence is defined in terms of severity and no distinction between it and confirmation is made.

---

[11]Mayo (2005, p. 99), where what counts as "extraordinary circumstances" is hinted at but not further defined.

[12]There is no need to repeat our demonstration that our likelihood ratio-based extended account of evidence can handle composite hypotheses. It is included as an Appendix, *A Note on Simple and Composite Hypotheses,* to Chap. 2.

[13]See Cox (2006), Lele (2004), Spanos (1999), and Dennis (2004), for examples.

To begin to focus these issues, consider the PAP smear test for cervical cancer in more detail. Although as we noted above, Mayo distinguishes between error-probabilities and likelihoods, it is convenient for the sake of our exposition to identify them. For one thing, it allows us to make a straightforward comparison with our own account. For another thing, and to the best of our knowledge, although Mayo warns us against reading her severity criterion as $Pr(D \mid H)$ is very low, she does not tell us in a formal and precise way how it is to be read. In any case, none of the criticisms we will make turn, so far as we can see, on temporarily construing the error-statistical position in terms of likelihoods. Again we make the same assumptions, viz., H is the hypothesis that an individual has a particular disease and $\sim H$ is the mutually exclusive and jointly exhaustive hypothesis that she does not, and D represents a positive screening result. We would like to know whether a test is in this context severe.

Cervical cancer is a disease for which the chance of containment and subsequent management as "chronic" is high if it is detected early. The PAP smear was until very recently a widely-used screening technique intended to detect a cancer that is as yet asymptomatic. An on-site proficiency test conducted in 1972, 1973, and 1978, assessed the competency of technicians who scan PAP smear slides for abnormalities.[14]

Overall, 16.25 % of the tests performed on the women with cancer resulted in false negative outcomes ($\sim D$). A false negative test occurs when the test of a woman who has cancer of the cervix incorrectly indicates that she does not. Therefore, in this study $Pr(\sim D \mid H) = 0.1625$. The other $100-16.25 = 83.75$ % of the women who had cervical cancer did test positive; as a result, $Pr(D \mid H) = 0.8375$.

Not all of the women who were tested actually suffered from cervical cancer. In fact, 18.64 % of the tests were false positive outcomes. This implies that $Pr(D \mid \sim H) = 0.1864$. The probability that the test results will be negative given that the individual tested does not have the disease is $Pr(\sim D \mid \sim H) = 1-0.1864$, or 0.8136.

On Mayo's account as we reconstruct it, T is a severe test of H given D just in case $Pr(D \mid H)$ is very high (condition 1) and $Pr(D \mid \sim H)$ is very low or (equivalently) $Pr(\sim D \mid \sim H)$ is very high (condition 2). As we have seen, in the PAP case, $Pr(D \mid H) = 0.8375$, which is high, $Pr(D \mid \sim H) = 0.1864$, which is low. It follows on the error-statistical account that the PAP smear is a severe test of the hypothesis that a given woman has cervical cancer.

On our account, which requires both evidence and confirmation for the test of a hypothesis to be severe, the PAP smear is not. We need to know, first, the $Pr(H \mid D)$, that is, the posterior probability that a person with a positive test result actually does have the disease. To apply Bayes Theorem, we need to know the prior probability of H in particular. The most reasonable (and in no way "subjective") $Pr(H)$ is the probability that a woman suffers from cervical cancer when randomly selected from the female population. In this case, it is simply the measure of the relative frequency of cervical cancer. One source reports that the rate of cases of

---

[14]See Pagano and Gauvreau (2000), pp. 137–138.

cervical cancer among women studied in 1983–1984 was 8.3 per 100,000.[15] That is, the data yield $Pr(H) = 0.000083$. It follows that $Pr(\sim H) = 0.999917$. Using the likelihoods given in the last two paragraphs and applying Bayes Theorem, we arrive at $Pr(H \mid D) = 0.000373$. So although the posterior probability of the hypothesis is larger than its prior, the hypothesis is only marginally confirmed. Hence on our account, T does not provide a severe test of H given D.

We need to know, second, whether D provides strong evidence for H. Since $Pr(D \mid H) = 0.8375$ and $Pr(D \mid \sim H) = 0.1864$, their ratio is 4.49. After we know that the individual tested has a positive PAP smear, her chances of having cervical cancer as against not having it have increased almost five-fold. While this might seem ominous, it is not, statistically speaking, a very large increase. It assuredly justifies a woman who receives a positive PAP result in taking more tests to make sure about her present state of health.[16] However, it does not justify a belief that she has the disease, nor does it provide very strong evidence that she does. The PAP smear satisfies neither of our criteria: it is simply not a severe test for the presence of cervical cancer.

In particular, the likelihood of the data given the presence or absence of cancer is not adequate to determine the posterior probabilities of the hypotheses. From a Bayesian perspective on confirmation, these posterior probabilities are important. To calculate them in the sort of diagnostic case we are taking as exemplary, we must also include information about the probability of an individual being afflicted when selected randomly from the population. In the PAP case, the prior probability of an individual's being afflicted is just 8.3 per 100,000 people. It is extremely low. At the same time, it plays a crucial role not simply in determining the posterior probability of an individual's having a disease, but also in saving us from generally counter-intuitive results.[17] In this case, if we were to *believe* on the basis of a positive result that a particular person had cervical cancer, we would be wrong the vast majority of the time. Evidence on the error-statistical account does not always provide us with reasons to believe that a hypothesis is true.

A number of philosophers have suggested that the error-statistical account of severe testing fails to take relative frequencies or base-rates into account.[18] Mayo's response is to accuse her critics, Bayesian and non-Bayesian alike, of committing

---

[15]*Ibid.*

[16]At this juncture, everyone confronts Royall's "what should I do?" question. Answering it involves utilities as well as probabilities, and lies outside the scope of the monograph.

[17]Achinstein (2001), pp. 134–136, sets out some very simple and schematic cases to make the same point, that neglect of prior probabilities, whether they are interpreted in an objective or subjective way, leads to counter-intuitive results. The list of other philosophers and statisticians, Bayesians and non-Bayesians alike, who have made the same criticism is long. See as an example of the Bayesian variety (Howson 1997). See also Bandyopadhyay and Brittan (2006) which was originally presented at the APA, Central Division in 2000.

[18]Again see Achinstein (2001), as well as Curd et al. (2012). p. 618. The importance of taking base-rates into account for a Bayesian account of confirmation was spelled out in our discussion of testing for tuberculosis in Chap. 2.

the "fallacy of instantiating probabilities."[19] This fallacy involves taking a relative frequency, of cases of cervical cancer in the general female population, say, and then attaching it to the hypothesis that a particular person has cervical cancer, viz., inferring the probability that Jane has cervical cancer is 0.0083 from the fact that 0.0083 percent of the female population has cervical cancer and that Jane is a female. That this sort of single-case inference from a statistical generalization is problematic has been understood among philosophers of science at least since Hempel's article, "Inductive Inconsistencies" (1960).[20] We are not going to fully unravel the problems here. But two comments about Mayo's support of her claim that the base-rate criticism itself rests on a "fallacy of probabilistic instantiation" are in order.

First, Mayo correctly draws attention to the problem of reference classes. It matters very much from what reference class a subject is drawn before we can infer a reasonable degree of belief that she has a particularly property. Hempel's examples are familiar. On the one hand, since Peterson[21] is a Swede and the proportion of Roman Catholic Swedes is less than 2 % we can infer that it is little likely that Peterson is a Roman Catholic. But this inductive inference needs to be set against another, that since Peterson made a pilgrimage to Lourdes and the proportion of those who make such a pilgrimage and are not Roman Catholics is less than 2 % we can infer that almost certainly Peterson is a Roman Catholic. True premises, inconsistent conclusions, we must conclude that both arguments are invalid. But this is not the end of the matter. As the smoking-lung cancer studies that will be a focus of the next chapter illustrate, it is possible in statistically careful ways, to take reference-class variation into account in both testing a thesis and inferring individual probabilities, e.g., that a male between the ages of 25 and 65 who has smoked two or more packs of cigarettes a day has an X probability of developing a cardio-vascular disease (notably lung cancer) and a Y probability of dying from it. Very sophisticated techniques have been developed in connection with epidemiological studies in particular to eliminate "biased" results. Of course, it is always possible that there is some heretofore unconsidered factor, something in the drinking water of all and only those who smoke and develop lung cancer that would greatly alter the probabilities.[22] The problem is the same one raised in our discussion of inherent dogmatism in Chap. 4, and is subject to the same comments: one would rigorously calculate a posterior probability representing an objective appraisal only if one could characterize all possible reference classes and assign prior probabilities to each. Second, clinicians working in the field[23] commonly

---

[19]See Mayo (2005, pp. 114ff.).

[20]In Hempel (1965).

[21]We've changed Hempel's "Petersen" to "Peterson." Petersen would be a Dane or Norwegian, not a Swede!

[22]Ibid.

[23]Schulzer (1994) is an especially clear discussion of the mistake often made by overlooking the notion of prior probability in analyzing diagnostic test results. See also Galen and Gambino (1975).

make diagnoses based in part on the relevant base-rates. To ignore them would not only lead to dramatically counter-intuitive results, it would invalidate a large swath of current practice. Better reasons than she provides are necessary to support a case for sweeping changes.[24] Nor is this a point of conflict between Bayesians and frequentists. Bayesians can view these diagnoses as posterior probabilities, while frequentists can view them as predictions conditional on a hypothesized reference class. The conflict that is exposed might be called a scope of inference conflict. A public health worker husbanding scarce resources and a patient whose life may be on the line probably should have very different opinions about which is the appropriate approach to take.

If a test satisfies our conditions of severity, it satisfies Mayo's as well.[25] On the other hand, as our examples make clear, a test that satisfies Mayo's conditions does not necessarily satisfy ours. Our conception is in this sense "more severe" than hers. It results from distinguishing and requiring both evidence and confirmation, at least understood as we have characterized them. It also better matches some of our scientific practice, particularly well, of course, when prior probabilities are determined objectively by way of relative frequencies as they are here.

Piecemeal Testing

The second main claim of the error-statistics methodology concerns piecemeal testing. A global theory passes a severe test by way of all of its component hypotheses passing what error-statistics understands as "severe tests."

Consider testing the General Theory of Relativity (GTR) to better understand how the error-statistical testing of a global theory proceeds. According to the GTR, gravity is a property of space-time geometry. More specifically, gravity is a space-time curvature. There are three *classical* tests for GTR, each of which tests a local hypothesis—H_{1E}, H_{2E}, and H_{3E}—severely. H_{1E} is the hypothesis that, as seen from the Earth, the precession of Mercury's orbit is 5600 s of arc per century. H_{2E} is the hypothesis that spectral lines from bodies with strong gravitational fields (e.g., the Sun and other stars) have longer frequencies than those that do not. Finally, H_{3E} is the hypothesis that gravitational fields alter the path of light. Since the explanation of Mercury's perihelion was the most important of all three tests in securing acceptance of the General Theory, we will discuss it first.

---

[24]In fact, Mayo admits the point Mayo (2005, p. 117): "Nor need we preclude the possibility of a statistical hypothesis having a legitimate frequentist prior. For example, a frequentist probability that Isaac is college-ready might refer to genetic and environmental factors that a high school student (from a specified population) is deficient—something we can scarcely cash out, much less compute" (*Ibid.*). But we cash out, indeed compute, such legitimate frequentist probabilities all the time in epidemiological studies.

[25]Insofar as it is possible to make a direct comparison between them since she avoids likelihoods and prior probabilities.

Perihelion is the epoch when a planet or comet is at its closest distance to the sun. The orbits of these heavenly bodies are not circular, but rather elliptical to varying degrees, and so there is a maximum distance, the aphelion, as well as a minimum, perihelion, distance from the Sun. It has been observed that Mercury's perihelion shifts over a very small distance. This shift of forty-three seconds of arc per century cannot be accounted for by classical gravitational theory; it predicts 5557 s of arc per century whereas the observed shift is 5600 s of arc. GTR is able to explain this residual 43 s.[26] The perihelion shift of Mercury is generally taken as a severe test of GTR and, as noted, very much helped to secure its acceptance. It clearly satisfies the error-statistical criteria: $Pr(D \mid H_{1E})$ is very high, i.e., close to 1, and the probability for its local Newtonian hypothesis, $Pr(D \mid H_{1N})$ is close to 0 where "D" is the observed shift of Mercury. Once we are able to show that H_{1E} has passed severely relative to a particular outcome and a specific test, then, following Mayo, we can say that "we have learned about one facet or one hypothesis of some more global theory such as GTR" (Mayo 1996, p. 190). To know whether the entire GTR has passed severely, and on the assumption that the three taken together comprise it, we need to know whether its other two local hypotheses, H_{2E} and H_{3E}, separately have passed their own severe tests.

The observational data for H_{2E} is that when light consisting of photons climbs out of a gravitational well it undergoes a time-dilation resulting in a red-shift, i.e., a shift in the spectrum of light emitted towards its red end. Again, $Pr(D \mid H_{2E})$ is very high, and for its alternative hypothesis H_{2N}, $Pr(D \mid H_{2N})$ is very low, in which case and on the error-statistical criteria, H_{2E} has passed a severe test. Similarly, detailed astronomical observations reveal that gravitational fields do alter the path of light, in which case the third local hypothesis, H_{3E}, also passes a severe test. Given these facts, error-statisticians are able to conclude that the global theory GTR has itself passed a severe test and therefore warrants acceptance.[27]

But the assumption on which this conclusion rests, viz., that if local hypotheses pass severe tests then the global theory passes a severe test and thus warrants acceptance, is questionable. In our view, this assumption incorporates a misuse of the conjunction law from probability theory.[28]

Suppose we want to submit some global theory, T, to a severe test. Following Mayo's guidance, we do so by submitting its component hypotheses, H_1, H_2, and H_3, to global hypotheses. Collectively they constitute a severe test of T if and only if $Pr(D \mid T)$ is high and $Pr(D \mid \sim T)$ is low, where D are the respective passings of their own severe tests by the three components. But the probability of the global theory passing this (meta-) test will be less than the probability of its local

---

[26]See Will (1993).

[27]Mayo (1996), p. 183.

[28]The lottery paradox is sometimes taken to show that conjunction poses problems for probabilistic epistemology. In the section on *Absolute and Incremental Confirmation* of Chap. 2 we have shown how this paradox can be resolved in terms of our dual conceptions of evidence and confirmation. It should also be mentioned that in the literature on classical statistics, there are discussions of how to collate p-values, but they are not germane to the present point.

hypotheses passing their own severe tests. In other words, since $\Pr(D \mid T)$ will be less than any of its conjuncts, it does not follow that if they are high, $\Pr(D \mid T)$ will also be high and the first criterion of error-statistical "severity" satisfied. We call this the "probability conjunction" error. It can be put more generally.[29]

Take a global theory, T, and several local hypotheses, $H_1, H_2, \ldots, H_i, \ldots$ and corresponding severe tests with outcomes $D_1, D_2, \ldots, D_i, \ldots$. These data are the passing or failing of each locally severe test. We know that for each i, $\Pr(D_i \mid H_i)$ is high, but not 1 (by the first error-statistical criterion for a severe test). By the rules of probability theory for conjunction, $\Pr(D_1 \& D_2 \& D_i \& \ldots \mid T)$ will be less than $\Pr(D_i \mid T)$ for all i. Hence if passing all the severe tests of its local hypotheses is taken as a severe test of the global theory, then $\Pr(D \mid H)$ will be less than the corresponding probability of any of the local hypotheses. Indeed, the more local hypotheses that pass severe tests, the less probable on the data would be the global theory that comprises them.[30] But this is counter-intuitive. The more local hypotheses that pass severe tests, the stronger should be the evidence for the global theory. The paradigm example of this is, of course, Newton's gravitational theory; as successful predictions on the basis of its component hypotheses piled up, among many others that tides and comets are alike subject to the same gravitational forces, the theory became more and more credible.

The "True-Model" Assumption

The third dimension of the error-statistical account of evidence we will discuss has to do with its incorporation of the "true-model" assumption, that strong evidence for a hypothesis provides us with good reasons for believing that it is true. It should be noted at the outset that there is room for another, possibly more charitable, construal of error-statistics, which fights shy of such a problematic "true-model" stance toward models/theories. Error-statistics often stresses the statistical adequacy of a model, which is different from the truth of a model. Statistical adequacy of a model concerns the empirical validity of the probabilistic assumptions underlying the estimated statistical model. So, instead of claiming that severe tests provide good reasons that model/theory are true or false, error-statisticians could make the more limited claim that models are statistically adequate or inadequate. We will eventually argue, however, that this construal has a price to pay for failing to account for

---

[29]We thank John G. Bennett for pointing out this objection to error-statistics.

[30]This objection is still more general and need not rely on the independence assumption that each local hypothesis be tested. On the error-statistical account, if we assume that each local hypothesis (three of them, for example) has passed a severe test, then each would constitute data D_1, D_2, and D_3 for the global theory, T. Then for any D_1, D_2, and D_3, $\Pr(D_1, D_2,$ and $D_3 \mid T) = \Pr(D_1 \mid T) \times \Pr(D_2 \mid D_1 \& T) \times \Pr(D_3 \mid D_1 \& D_2 \mid T)$ for the probability of the conjunction for the global theory, but this probability for the global theory will still be less than $\Pr(D_1 \mid T)$ because each of the other terms is less than 1, as required by the inductive nature of the error-statistical account.

causal claims made in global scientific theories. Moreover, the more limited claim does not cohere very well with what we might call the "logic" of Mayo's argument, or with the many occasions on which she urges the less limited "true-model" assumption.

We can begin with her words: "passing … severe tests warrants accepting H" (Mayo 1996, p. 183). Or again, "When a test is sufficiently severe, … the passing result may be said to be a *good indication* (or a good ground for) H," and again, "evidence indicates that H is true" (*Ibid.*, p. 186), although with the same breath (*Ibid.*, p. 186), she sometimes takes a model adequacy stance: "Within a canonical experimental test, the truth of H means that H gives an approximately correct description of the procedure generating experimental outcomes." The general drift of her remarks is to embrace the "true-model" assumption. Her recent reference to Glymour's long-time move to relate a theory's ability to provide a good explanation with the theory being true is still another example. She writes, "Like Glymour, I deny that what matters is explanatory power as *opposed* to truth—truth does matter" (Mayo and Spanos 2010, p. 354).[31]

Moreover, there are at least three reasons why Mayo must make something like the "true-model" assumption. We will discuss them in order of increasing importance.

The first reason has to do with the way in which she parts company with Popper. On her reading,[32] the latter was committed to our comparative likelihood account of evidence. Mayo concludes correctly that on this account, evidence does not provide us with good reasons to believe that a hypothesis is true. Indeed, this is one of our main themes. From this she draws the further conclusion that Popper had to settle for falsification and the correlative notion of "corroboration," that the most we can claim for a theory is that it has not yet been falsified. But, in contrast, it follows from the error-statistical approach that "when a hypothesis H has passed a highly severe test we can infer that the data **x** provide good evidence for the correctness of H."

The second reason why error-statistics must embrace the "true-model" distinction is simply an elaboration of the first. Mayo conflates evidence and confirmation, implicitly maintaining that the second reduces to the first. But the first on our (Popperian) comparative likelihood account of it does not do what any account of confirmation must, viz., provide good reasons for believing that the hypothesis confirmed is true. So she rejects our account of evidence[33] and replaces it with an error-statistical account which, in her view, *does* provide us with appropriate good reasons. But this is to say that she must embrace the "true-model" assumption. Our view, of course, is that evidence and confirmation must be distinguished. Confirmation but not evidence makes this assumption. We have also argued that the assumption provides ready ammunition for those who want to attack the pre-eminent

---

[31]See Mayo and Spanos (2010, p. 354).

[32]Whether this is a correct reading of Popper does not concern us. See Mayo (2005) for the material to support our reading of her position.

[33]This is not the only reason she does so, but it is "crucial" (*Ibid.*, p. 105).

credibility of scientific methods, in particular the gathering of objective evidence for one theory or model as against another.

The third and most important reason that error-statistics assumes that some models are true has to do with a deep tension in its account of the results of severe testing. On the one hand, the error-statistical account presupposes that you can test, and thus gather evidence for, a hypothesis or model in isolation. It is in this way that a severe test constitutes an acceptance or rejection procedure for the hypothesis; it supplies good reasons for believing that the hypothesis is true or false (or, synonymously, "correct" or "incorrect"). Some of our own examples, in which the probability of the data on a hypothesis and its negation are contrasted, might have suggested as much. But the testing of a hypothesis in isolation presumes not only that it and its negation are mutually exclusive but that they are jointly exhaustive. In the real world of science, this second presumption is rarely true. The fact of the matter is that there are a variety of ways in which a hypothesis can be false. It might be, for example, that a hypothesis H is tested severely and successfully against its negation $\sim H$, but that it cannot be so tested against another hypothesis H'. This comes to saying that although $\Pr(D \mid H)$ might be high, $\Pr(D \mid H')$ might also be high. Although in reconstructed cases, one theory is pitted against another, it is much more typical to have a spectrum of theories under consideration, not just Copernicus' and Ptolemy's (understood in an overly-simplified way as mutually-exclusive and jointly-exhaustive kinematic descriptions of the solar system), but Copernicus', Ptolemy's, Brahe's, and Kepler's; as in the usual whodunit, there is not one suspect or perhaps two, but several. Mayo occasionally chides Royall for allowing too many alternatives. The criticism is more pointed in reverse. Mayo allows too few alternatives to be an accurate representation of actual scientific practice. She often stresses her desire to reflect actual scientific practice, but here she is flying in the face of a long tradition of practicing scientists—from Chamberlain (1897) to Platt (1964) to Burnham and Anderson (2002), for a century and a quarter multiple models have been held to be a good thing. Multiple models ask the question "how can I be wrong?", while a single model, even if carefully iterated, only asks "how can I be made to seem right?" or "what errors need to be corrected?" Chamberlin's key paper on *The Method of multiple working hypotheses* put it this way: "With this method the dangers of parental affection for a favorite theory can be circumvented." What he has in mind is the ever-present danger of bias. Equally important in our view is that evidence has real bite only when it serves to distinguish between multiple models. Human-caused and ocean-temperature-caused global warming are not simply mutually-exclusive and jointly-exhaustive alternatives; evidence for and against them may be gathered in a genuinely comparative context. The same is true of the alcohol- and smoking- lung cancer hypotheses. Evidence is gathered to separate and compare them, prior to any assessment of the "severity" of the tests to which they are submitted, even as it prompts the search for new and better models to test. But this fact implies that the "true-model" assumption is inappropriately incorporated into an account of evidence. This implication has to be set beside the error-statistical claim that severe testing provides good reasons to think that particular hypotheses are true.

Although the argument undermining Mayo's "multiple-model" criticism of our own Royall-derived evidential account is rather technical, and therefore put in an Appendix to this chapter, its conclusions can be summarized briefly. First, in both evidence and severe testing, there is model comparison. It is explicit in evidence and implicit in severe testing. Second, in both evidence and severe testing, there is a statistical test. It is implicit in evidence and explicit in severe testing. Third, in both evidence and severe testing, there is estimation of post-data error probabilities and is explicit in severe testing. The kinds of differences alleged between the two accounts do not exist.

As we mentioned earlier in this section, error-statisticians sometimes suggest a more limited claim, that severe tests demonstrate only the statistical adequacy of hypotheses, and do not also give us good reasons to believe that they are true. It is not simply that the more limited claim is sometimes made. It would also be forced on them if, as we have just contended and argue in more detail in the Appendix to this chapter, their account of severe testing must acknowledge, however implicitly, a connection between the concept of evidence and multiple models. For once multiple models are admitted, as we think they must be on any adequate account of *evidence* as it is correctly understood in scientific practice, then, as is also argued, the "true-model" assumption must be abandoned.

Statistical adequacy deals with the empirical validity of the statistical assumptions of a model. Philosophically speaking, a model is statistically adequate just in case it is able to describe data up to a desired level of precision. The error-statistical account of severity is one step in determining the statistical adequacy of a hypothesis. Error-statisticians like Mayo couple their account of severity with misspecification of a hypothesis in order to provide a full-blown test of its statistical adequacy. In regression analysis, for example, the model specification consists of selecting an appropriate functional form for the model and choosing which variables to include in it. It can also arise from various considerations, among them omitting a variable that might be related to both dependent variables and one or more of the independent variables is one of them. Model misspecification could also stem from an irrelevant variable in the model. However, the construal of a model/theory as statistically adequate instead of true has its own short-comings, which we will proceed to elaborate.

Error-statisticians seem to be confronted with two options. Either they could opt for the true-model assumption, which in the case of global theories involves having good reasons to believe that the causal mechanism postulated by such theories explains the phenomena described by its local hypotheses or make the more limited claim that we have good reason to believe that the models are statistically adequate. On either option, error-statistics suffers from defects. If error-statistics holds the true-model assumption concerning global scientific theories, then, since we may assume that truth and causal explanation are connected in the way that Glymour, Mayo, and many others insist, it is able to make truth-claims about causal explanations. Global theories as commonly understood do not simply describe what is the case but explain why.[34] But in the process error-statisticians might seem to have committed the fallacy of affirming the consequent.

P1. If a theory/model is true, then the value of the test-statistic should be closer to the value derivable from the model in question

P2. The value of the test-statistic is closer to the value derivable from the model in question
     Therefore, the theory/model is true.

The erroneous nature of this inference is obvious. Statistical literature is replete with cases in which infinitely many competing models may satisfy premise 2. Yet we can't conclude from this that they are all equally true models. Making the true—model assumption is bound up with the explanatory character of global scientific theories. But making this assumption, as we have just seen, gives rise to an invalid inference. So, error-statisticians could instead retreat to the claim that models tested successfully are no more than statistically adequate. This move avoids the error by claiming that the model in question describes the data adequately, but fails to make provision for the causal-explanatory character of global scientific theories. Therefore, error-statistics is confronted with a dilemma. Either embrace the true-model or the statistical-adequacy assumption. If the former, then it is able to provide causal explanations of the phenomena, the widely agreed-upon aim of most global scientific theories. But, in the process error-statistics commits the fallacy of affirming the consequent. By contrast, if it gives up on the true-model assumption then it does not commit the fallacy, yet it loses the possibility of providing causal explanations. Given the link between truth and explanation, one cannot have both mere statistical adequacy and causes.

None of this poses any threat to our comparative likelihood account of evidence. We will make three comments. First, we are able to make inferences about causal mechanisms by comparing two causal models to see which one of them is more supported by the data. Second, we do not claim on the basis of the evidence that either of the causal models is true, because there could be a better model which we are yet to consider which is more supported by the data than the two considered so far. Third, we make a *local* causal inference about the relative evidential strength of those two models because we make *inferences* regarding the comparative strengths of *only two* models at a time, and *not* for *all* causal models. Whether one or the other of the models is confirmed is another matter. Confirmation is to be distinguished from evidence. Unlike Mayo, we do not try to have it both ways.

---

[34]Newton's theory explains why the orbits of the planets are (more or less) elliptical, Einstein's (general) theory explains why light bends in the vicinity of large masses such as the Sun, the examples are both numerous and familiar.

Further Reflections

Classical error-statistics provides researchers with a variety of useful tools, confidence intervals and significance tests among them, and rightly focuses attention on error and reliability in the assessment of data, not to mention the corollary necessity of random sampling and other statistical techniques. But from our perspective, Mayo's attempt to fashion these tools into an account of Popper-like severe testing suffers from at least two problems.

First, her account is both too weak and too strong. It is too weak because it is open to counter-examples of the PAP smear-cervical cancer variety. At the same time, it is too strong because it precludes the necessity of assessing the evidence for a variety of hypotheses with respect to the same set of data. Like Popper, she thinks that hypotheses can be tested in isolation, but only sometimes is this possible, and even more rarely is it satisfactory.

Second, and unlike Popper, Mayo often gives the impression that passage of a severe test provides us with a good reason to believe that it is true. That is to say, instead of falsification, which relies on a straightforward application of *modus tollens*, error-statistics seems to be committed to a version of *modus ponens*. If local hypotheses *H1, H2,* and *H3* pass severely then the global theory *T* which comprises them also passes severely. Given that the antecedent of the conditional is true, the conclusion follows. We have argued that the first premise is false, consequently, the argument is unsound. Although each local hypothesis, *H1, H2,* and *H3* passes severely, it does not follow that *T* will thereby pass severely, as this inference violates the probability conjunction rule. Perhaps for this reason or some other, error-statisticians seem increasingly to play down the connection between piecemeal hypothesis and global theory testing. "The growth of knowledge has not to do with replacing or confirming or 'rationally accepting' large-scale theories, but with testing *specific hypotheses* in such a way that there is a good chance of learning something-whatever theory it winds up as part of" (Mayo 2010, p. 28, emphasis ours). This more modest version of the error-statistical account, that it is concerned solely with testing local hypotheses and controlling their errors (and only as a kind of corollary with the appraisal of theories of which they may happen to form part), avoids the probability—conjunction and possibly other similar sorts of objections. But it pays a price in doing so and it suggests a very misleading relationship between global theories and their component hypotheses.

The price paid is that it no longer provides an alternative account of theory-acceptance or rejection, on the same scale as the Popperian or Kuhnian paradigms, or for that matter the other less iconic accounts surveyed in this monograph. The status of an error-statistician if solely interested in testing local hypotheses might now seem comparable to a mechanic who can identify problems and make repairs to one part of a boat at sea or another, but can't promise (to any degree of certainty) that the entire boat won't be capsized.

The more modest approach suggested at times is also misleading in two respects. Consider this general and relatively recent error-statistical claim:

A severity assessment is not threatened by alternatives at 'higher levels.

If two rival theories, T_1 and T_2, say the same thing with respect to the effect or hypotheses *H* being tested by experimental test E...then T1 and T2 are *not rivals* with respect to experiment E. Thus, *a severity assessment can remain stable through changes in "higher level theories"* or answers to different questions. For example, the severity with which a parameter is determined may remain despite changing interpretations about the cause of the effect measured (Mayo 2010, p. 36). On the one hand, we are invited to infer that if two rival theories do not say "the same thing" with respect to the hypothesis being tested then they are rivals with respect to the test, in which case there are grounds for preferring one to the other, grounds that presumably error-statistics can make clear. In this case, it would seem that Mayo believes error-statistics provides the means for global as well as local theory assessment. As statistical adequacy is the only inferential tool that error statisticians have, it seems inevitable that the rivals T1 and T2 will be compared on the basis of it alone. Lele (2004) points out that evidence functions are the differences of statistical adequacy measures. Thus, it seems that the desire to assess rival theories will transform an error statistician into an evidentialist, perhaps unknowingly.

On the other hand, and depending on what the vague "saying the same thing" is intended to mean, this claim suggests a rather naïve view of the relationship between global theory and local hypothesis. At the time when Newton began his work on gravitational attraction, there were other ways of calculating planetary orbits besides Kepler's, of roughly comparable accuracy. All were elliptical. Presumably they would have passed the same severity assessment with respect to the data. But Newton's "higher level" theory was able to show that elliptical trajectories were not simply mathematically tractable, but required dynamically.[35] It provided a well-supported causal explanation. Perhaps more important for present purposes, with the aid of his theory corrections were made in the determination of these trajectories, which in turn led to modifications of the theories, which in turn led to more accurate determinations, a dialogue between global theory and local hypothesis which seems to be left out of the rather casual claim that local severity assessments are left untouched by the theoretical contexts in which they are made. Like Popper, we believe that assessments of global theories must be made, if for no other reason than that just mentioned, that a sharp separation of them from their constituent hypotheses is not possible, not simply because of the causal and explanatory connections of the former to the latter, but also because of the parameter adjustments that the former allows us to make in the latter. Like him, we give up the "true-model" assumption. Unlike Popper, however, we think that evidence can nonetheless provide us with approximations to the truth, as well as a way of measuring their distances from it. We think there is something called "the truth" or "reality." But, what we deny is the true-model assumption. As we will see more

---

[35]As Newton famously wrote Halley, "Kepler knew the Orb to be not circular but oval, and guest it to be elliptical," whereas I have now shown why it must be so.

clearly in the next chapter, his view that hypotheses can be falsified by deducing consequences which do not match what we in fact observe, and therefore that we can dispense with probability theory and its attendant measures of confirmation and evidence, rests on untenable assumptions concerning the possibility of testing hypotheses in isolation.

Appendix

A Note on Severe Testing, Evidentialism, and Multiple Models

Mayo distinguishes her severe testing account from Royall's evidential account along three axes, first that she is concerned with the evaluation of a single hypothesis and not with the comparison of hypotheses, second that tests with error rates are superior to likelihood ratio evidence, and third that severity involves post-data error rate calculations which she thinks LR evidence does not. Let us try and understand severity a little better to see if her critique is valid.

Mayo and Spanos (2006) say that:

A statistical hypothesis H passes a **severe test** T with data x_0 if,
(S-1) x_0 agrees with H and
(S-2) with very high probability, test T would have produced a result that accords *less* well with H than x_0 does, if H were false.

This is a general form which can't be applied or assessed without two things: (1) a method to determine if x_0 agrees with H and just as importantly (2) a method to determine the distribution of x_0 if were false.

While requirement 1 can be met with Fisherian tests that do not express an alternative model, requirement 2 can only be met if there is an explicit alternative model or family of models. To repeat, severity, Mayo's key inferential innovation, can only be calculated under an explicit alternative model. Mayo's language suppresses recognition of the alternative by folding it into a "test of H", but it must be present nonetheless.

To further quote from Mayo and Spanos (2006):

$$SEV(\mu \leq \mu_1) = P(d(X) > d(x_0); \mu \leq \mu_1 false) = P(d(X) > d(x_0); \mu > \mu_1).$$

The severity calculation given above is the second step in a severe test. The first step is the evaluation of a test of H of size α, $T(\alpha)$. This evaluation will involve the null model μ_0, whether the test is a Neyman-Pearson test or a Fisherian test. Thus, to conduct a severe test there must be at least two models which are being compared. This deflects her first critique of Royall. The examples that Mayo uses tend to be comparison of the means of normal distributions. She uses a function of $\bar{x}$ as her test statistic, $d(x_0)$. As $\bar{x}$ is a sufficient statistic for the mean of a normal distribution, a corollary to the Neyman-Pearson theorem indicates that her test is equivalent to a test using the likelihood ratio as a test statistic. Such a test can be written as $LR_{1,0} < k_\alpha$, where k_α is selected so that the test will have size α.[36] In fact, if there is a "best test," it will be, or be equivalent to a likelihood ratio test, with "best test" being defined in terms of lowest probability of type II error, exactly Mayo's currency. If there isn't a best test, but only a uniformly most powerful test, it will be, or be equivalent to, a likelihood ratio test. If there isn't a uniformly most powerful test, the likelihood ratio will still be a good test.[37] Neyman-Pearson tests are isomorphic with likelihood ratio evidential comparisons with a given test size specifying a strength of evidence and vice versa. This deflects Mayo's second criticism.

Another thing to note is that the severity expression above can't be evaluated without placing a prior on μ. Mayo and Spanos actually evaluate a lower bound for severity given by:

$$SEV(\mu \leq \mu_1) > P\big(d(X) > d(x_0); \mu = \mu_1\big).$$

That is severity is an error probability bound evaluated at the alternative model. Taper and Lele (2011) point out that the probability of obtaining misleading evidence as strong as the observed evidence (M_L) has an upper bound of $1/LR_{1,0}$. Severity and post-data (local) probability of misleading evidence are compliments of one another (i.e. $M_L = 1 - Sev$),[38] and thus contain the same information. A post data error bound calculation is implicit in every likelihood ratio evidential assessment. The calculation needs to be made explicit in severe testing because of the granularity of dichotomous test in the first step of severe testing. This deflects Mayo's third critique.

The concept of misleading evidence will be discussed further in Chap. 8.

References

Achinstein, P. (2001). *The book of evidence*. Oxford: Oxford University Press.
Bandyopadhyay, P., & Brittan, G. (2006). Acceptance, evidence, and severity. *Synthèse, 148,* 259–293.
Burnham, K., & Anderson, D. (2002). *Model selection and multi-model inference: A practical information-theoretic approach* (2nd ed.). New York: Springer.
Casella, G., & Berger, R. (1990). *Statistical inference*. Garden Grove: Duxbury Press.
Chamberlain, T. (1897). The method of multiple working hypotheses. *Journal of Geology, 5,* 837–848.
Cox, D. (2006). *Principles of statistical inference*. Cambridge: University of Cambridge Press.
Curd, M., Cover, J., & Pincock, C. (Eds.). (2012). *Philosophy of science: The critical issues*. New York: W.W. Norton.

---

[36]See Casella and Berger (1990), Corollary 8.3.1.

[37]See Hogg (1978), Chap. 7 on "Statistical Hypotheses".

[38]With possibly some minor discrepancy due the bounds calculation.

Dennis, B. (2004). Statistics and the Scientific Method in Ecology. In M. Taper & S. Lele (Eds.), *The nature of scientific evidence: Statistical, philosophical, and empirical considerations*. Chicago: University of Chicago Press.

Fisher, R. (1930). Inverse probability. *Proceedings of the Cambridge Philosophical Society, 26*, 528–535.

Galen, R., & Gambino, R. (1975). *Beyond normality: The predictive value and efficiency of medical diagnoses*. New York: Wiley.

Hempel, C. (1965). *Aspects of scientific explanation*. New York: Free Press.

Hogg, R. (1978). *Introduction to mathematical sciences* (4th ed.). New York: Macmillan.

Howson, C. (1997). Error probabilities in error. *Philosophy of Science, 64*(Supplement), 185–194.

Lele, S. (2004). Evidence function and the optimality of the law of likelihood. In (Taper and Lele, 2004).

Mayo, D. (1996). *Error and the growth of experimental knowledge*. Chicago: University of Chicago Press.

Mayo, D. (1997a). Duhem's problem, The Bayesian way, and error statistics, or 'what's belief got to do with it?' *Philosophy of Science, 64*, 222–224.

Mayo, D. (1997b). Error statistics and learning from error: making a virtue of necessity. *Philosophy of Science, 64*(Supplement), 195–212.

Mayo, D. (2005). Evidence as passing severe tests: highly probable versus highly probed tests. In (Achinstein, 2005), pp. 95–128.

Mayo, D. G., & Spanos, A. (2006). Severe testing as a basic concept in a Neyman-Pearson philosophy of induction. *British Journal for the Philosophy of Science, 57*, 323–357.

Mayo, D., & Spanos, A. (Eds.). (2010). *Error and Inference*. Cambridge: Cambridge University Press.

Neyman, J. (1976). Tests of statistical hypotheses and their use in studies of natural phenomena. *Communications in Statistics—Theory and Methods, 5*, 737–751.

Pagano, M., & Gauvreau, K. (2000). *Principles of biostatistics*. Garden Grove, CA: Duxbury.

Pearson, E. (1955). Statistical concepts and their relation to reality. *Journal of the Royal Statistical Society, 17*, 204–207.

Platt, J. (1964). Strong inference. *Science, 146*, 347–353.

Popper, K. (1959). *The logic of scientific discovery*. New York: Basic Books.

Schulzer, M. (1994). Diagnostic tests: A statistical review. *Muscle and Nerve, 7*, 815–819.

Sellke, T., Bayam, M., & Berger, J. (2001). Calibration of p values for testing precise null hypotheses. *The American Statistician, 55*(1), 62–71.

Spanos, A. (1999). *Probability theory and statistical inference*. Cambridge: Cambridge University Press.

Taper, M. & Lele, S. (2011). Evidence, evidence functions, and error-probabilities. In P. Bandyopadhyay & M. Forster (Eds.), *Handbook of Statistics*. Amsterdam: Elsevier, North-Holland.

Will, C. (1993). *Was Einstein right? Putting general relativity to the test* (2nd ed.). New York: Basic Books.

Chapter 7
Selective Confirmation, Bootstrapping, and Theoretical Constants

Abstract Clark Glymour's "bootstrap" account of confirmation rightly stresses the importance of selective confirmation of individual hypotheses, on the one hand, and the determination of theoretical constants, on the other. But in our view it is marred by a failure to deal with the problem of confounding, illustrated by the demonstration of a causal link between smoking and lung cancer, and by the apparent circularity of bootstrap testing (which is distinguished from statistical bootstrapping). Glymour's proper insistence on a variety of evidence is built into our account of evidence and not added on as a way of handling the apparent circularity in his account. We discuss and dissolve his well-known charge against Bayesian theories of confirmation, that they lead to the paradox of "old evidence," in Chap. 9.

Keywords Bootstrapping · Selective confirmation · Theoretical constants · Confounding · Bootstrap circularity · Variety of evidence

Setting the Context

Clark Glymour's "bootstrap" account of confirmation[1] is like Deborah Mayo's in that they both eschew anything like belief probabilities and disregard the central evidential role we assign to likelihoods and their ratios. As we have just seen, Mayo's main reason for resistance has to do with the subjectivity introduced into scientific methodology by the indispensability of prior probabilities in Bayesian inference, although as we noted early on, she also thinks the use of likelihoods and their ratios in an account of evidence is problem-ridden. As for Glymour, the resistance has also to do with conventional problems surrounding the determination and distribution of priors,[2] but more especially with what he thinks is the in-principle dispensability of all probabilities in the analysis of paradigm examples of inference in the history of science, and also with his apparent demonstration that Bayesian confirmation leads to

---

[1]See Glymour (1980).
[2]See both Glymour early (1980) and late (2010).

© The Author(s) 2016
P.S. Bandyopadhyay et al., *Belief, Evidence, and Uncertainty*,
Philosophy of Science, DOI 10.1007/978-3-319-27772-1_7

paradox, notably in the form of the "old evidence problem".[3] Moreover, both the error-statistical and "bootstrap" accounts fail to recognize any distinction between confirmation and evidence. Beyond that, the differences between them are significant. Mayo's account is quantitative, Glymour's qualitative for starters.[4] So far as he is concerned, statistics plays no role in theory-testing.

But more important, they are animated by very different concerns. Mayo wants to provide an account of severe testing that both mirrors crucial aspects of scientific practice and explains why; on her error-statistical account, there is evidence for and hence good reason to believe individual hypotheses that have passed severe tests with respect to the data. But for Glymour, the issue is not severity but selectivity; in his well-known view, the actual practice of science is concerned mainly with establishing the relevance of particular observational, especially measurable, data to an individual hypothesis.[5] The task of confirmation theory shifts to making clear the general structure of the arguments that do so. This in turn requires showing how, in a rigorous way, observation selectively supports theory. Glymour does it by adding further constraints to existing accounts of the confirmation relation and proposing a strategy for proceeding, using one or the other of these accounts, to confirm or disconfirm hypotheses.

Selective Confirmation

The problematic character of selective confirmation was first emphasized by the French physicist and philosopher, Pierre Duhem.[6] Duhem pointed out that hypotheses in physics are never tested in isolation, but only in conjunction with other so-called auxiliary hypotheses. The use of telescopes to confirm Newton's theory of gravitation, for example, depends on assuming the correctness of a particular optical theory, among many other such theories, and thus on the acromatic corrections in astronomical observations that it makes possible. On the traditional hypothetico-deductive model that Duhem accepts, to test a hypothesis is to draw observational consequences or predictions from it. If the consequences are verified in the course of our experiments, then the hypothesis is to that extent confirmed, if not, then it is falsified. Duhem was not particularly worried about the fact that a hypothesis being tested must share the credit with all of the auxiliary hypotheses (in his view, ultimately all of physics) necessary to derive observational consequences

---

[3]To be discussed in Chap. 9.
[4]See Douven and Meijs (2006), for an elaboration of Glymour's account as a quantitative theory of confirmation.
[5]Although his recent work on causal inference has shifted his focus to severe testing in which "explanation and explanatory virtues are essential to produce testable hypotheses …" Glymour (2010, p. 349), which, we noted in the last chapter, Mayo has incorporated in her own account of severe testing.
[6]Duhem (1954).

from it, although Glymour is.[7] Duhem's worry concerned the attribution of blame, the apparent failure of a negative result to single out one of a number of conjoined hypotheses as false. "The only thing the experiment teaches us is that among the propositions used to predict the phenomenon and to establish whether it be produced, there is at least one error, but where this error lies is just what it does not tell us".[8]

It is difficult to overstate the influence of this claim on philosophy of science in the 20th century.[9] "Since logic does not determine with strict precision the time when an inadequate hypothesis should give way to a more fruitful assumption,"[10] the door was open to a sweeping new view of theory change on which it was in some fundamental sense illogical or "irrational".[11]

---

[7]Under the heading of "irrelevant conjunction." If a hypothesis is confirmed when a successful observational prediction is derived from it, then so too is any hypothesis, relevant or not, with which it can be consistently conjoined, a matter of elementary logic. If H is confirmed by D, so too are H and H', however irrelevant H' might be to the derivation of D. But we do not have to worry about undeserved credit within our framework, first, because what Duhem refers to as "the common sense of the working scientist" (who ignores arbitrary and therefore irrelevant hypotheses) is built into the prior probability on which the confirmation of hypotheses in part depends, second, because within our framework and in the most common cases, observational predictions per se are not "derived" from hypotheses; what is derived from hypotheses are the probabilities or probability distributions of observations expected under hypotheses, third, because irrelevant hypotheses can be identified *ab initio*: if $Pr(H \mid D) = Prob(H \& H' \mid D)$, then H' is "irrelevant," i.e., not confirmed by D. So much, it would seem, for the confirmation of irrelevant hypotheses. But in an e-mail communication, Glymour contends that our likelihood-based account of evidence also leads to the irrelevant conjunction problem. If data provide evidence for one hypothesis against a competitor, then they also provide evidence for the first hypothesis + an irrelevant proposition, e.g., that the Pope is infallible. Consider H_1, the hypothesis that Jones has tuberculosis, and H_2, the hypothesis that H_1 is false. Let's assume that Jones's X-ray is positive, and that $Pr(D \mid H_1)$ is very high, and $Pr(D \mid H_2)$ is very low. In such a case, the LR value for H_1 over H_2 is very high, in which case the data provide strong evidence for the hypothesis that Jones has tuberculosis and the Pope is infallible as against the hypothesis that Jones does not have tuberculosis. We grant the objection, but it has little sting so far as scientific practice is concerned. The likelihood account of evidence is to be applied within a conventional statistical framework. In this framework, the hypothesis and the hypothesis + irrelevant conjunct are not an estimable pair. H_1 and H_2 are not estimable because, given the data in hand, there is no way in which to distinguish between them. They would be estimable if we were to gather data about Papal infallibility (and not simply more positive X-rays), but with the data in hand this point, they cannot be distinguished. As such, and as a matter of scientific and statistical practice, what we have dubbed the evidential condition cannot (yet) be applied to them.

[8]Duhem (1954, p. 185).

[9]For a notable example, it appears to undermine Karl Popper's claim that whereas confirming consequences of a hypothesis, no matter how many, never prove that it is true, a single disconfirming evidence entails, as a matter of logic, that it is false As soon as two or more hypotheses are needed to deduce the consequences, as Duhem argues, disconfirming results can never show as a matter of elementary logic which one is false.

[10]Duhem (1954, p. 218).

[11]See Kuhn (1970), and the discussion of the "Kuhnian" argument in Chap. 1 against the possibility of the sort of descriptive-normative account that we and the other authors discussed in Part II of the monograph are trying to provide.

"Bootstraps"

Glymour wants to close this door. In his view, the paradoxical claim that neither selective confirmation nor disconfirmation is possible in principle runs up against the widespread intuition that both are routine, that individual hypotheses are accepted or rejected on the basis of experiment. The task is to show precisely how it is possible. The key to the account is that hypotheses are confirmed or disconfirmed by verifying positive or negative instances of them, and instances are verified by computing or determining the value of *every* quantity or property they contain. Those quantities or properties not directly measured in, or determined by, experimental observations may be calculated by using auxiliary hypotheses, including the very hypothesis from which the instance was derived. Quantities that are neither measurable nor computable are indeterminate, and hypotheses containing them cannot be tested; they are for this reason empirically meaningless and therefore "irrelevant". Confirmation, then, is a three-termed relation linking a piece of evidence E[12] to a hypothesis H by way of a background theory T, with the provisos that other evidence might have led to the derivation of a negative instance and that H might be included in T. It is this last feature, that a hypothesis may be assumed in the course of testing it, that a hypothesis can and often does support (or "lift") itself by its own bootstraps, that has come to define the position.

The testing strategy is straightforward. An experiment to test a hypothesis is carried out against a particular theoretical background. The experiment involves making a series of measurements. Some quantities that the hypothesis ascribes can be determined by the experiment; others cannot. Those that cannot (often some theoretical constant) are computed by means of other hypotheses of the theory (or even the original hypothesis itself) and available empirical generalizations. The computed values confirm or disconfirm the hypothesis by way of providing positive or negative instances of it.

In the simplest case, a hypothesis may be used to test itself and, in more complex cases, hypotheses belonging to the same theoretical framework can be made mutually supportive. Thus the basic gas law, $PV = kT$, can be used to test itself simply by conducting two or more experiments in which values for P (pressure), V (volume), and T (temperature) are obtained for instances of the generalization, and then seeing whether the inferred values for k, a constant that cannot be measured directly, match. If they do, the law is confirmed by the experiments, if not, it is disconfirmed. The crucial points are that a value for k has to be determined before the hypothesis can be tested, since a hypothesis can be tested only when *all* its values are determined, and that one, although not the only, way to do this is to use the hypothesis itself. The apparent circularity involved is not vicious, for the second set of measurements *could* have led to a negative result, that is values such that

---

[12]As with every other author discussed in Part II of the monograph, Glymour does not distinguish between "data," D, and "evidence," E, a consequence of their all conflating confirmation and evidence.

$PV \neq kT$. The "bootstrap" strategy requires that in every case such negative results are at least possible.

In more complex cases, one hypothesis in a theory is used to test another hypothesis in the same theory by deriving, in a non-trivial way and often with the aid of one or more empirical generalizations, the value for a constant which both contain, or data which agree with each. Thus, Newton used Kepler's second law, that a line segment joining a planet and the Sun sweeps out equal areas during equal intervals of time, together with his own second law of motion which allows us to measure force in terms of mass and acceleration, to derive instances of the law of universal gravitation for the planets and their satellites.[13] We "pull ourselves up by our bootstraps" because we use certain parts of a theory to obtain evidence for other parts of the same theory without having to assume "auxiliary hypotheses" from other theories. Testing in this way can be confined to particular theoretical contexts; it does not inevitably implicate, as Duhem insisted, other more distant hypotheses.

A somewhat more complex example is similarly drawn from Newton's gravitational theory.[14] Newton used his first two laws to derive Kepler's third law, that the square of the orbital period of a planet is proportional to the cube of its orbital radius. Using terrestrial experiments as evidence together with his own third law that for every force there is an equal and opposite force, Newton was able to deduce that a planet must exert an equal and oppositely directed force on the Sun that the Sun directs on the planet. Newton then argues that these forces must depend on the masses of both objects. In this way Newton confirmed the law of universal gravitation which asserts that all bodies exert an inverse square attractive force on one another. Newton used bits of his own theory of motion (first and second laws) to derive an instance of the same theory (law of universal gravitation) with the help of terrestrial observations.

In *Theory and Evidence*, Glymour writes, "[t]he key question, ..., is how it is possible for evidence... to test, support, and confirm hypotheses [without invoking hypotheses from other theories]".[15] He answers, "[bootstrap confirmation] illustrate [s] the fashion in which the testing procedure localizes confirmation, permitting one to say that particular pieces of evidence bear on certain hypotheses, but not on others, in a complex theory" (p. 106).

On Glymour's account of selective confirmation, it should be underlined, probabilistic reasoning is in principle not needed (or at any rate marginal), and,

---

[13]In this and other examples in *Theory and Evidence*, Glymour uses Hempel's "positive instance" account of "confirming evidence" (to be described in the Chap. 9) as paradigm, but emphasizes that the use of one part of a theory to test other parts of it can and must be added to other traditional accounts. He does not assume the correctness of, still less defend, the positive instance account, although he does think that it provides a very good reconstruction of Newton's method in *Principia*.

[14]We have very much simplified both examples. One of the many virtues of Glymour's account is the detailed way in which he tries to align Newton's own reasoning with it. At the same time, the alignment can be criticized in the same detailed way, as by Laymon (1983).

[15]Glymour (1980, p. 110).

worse, as we shall see in Chap. 9, it leads in his view to a fundamental paradox, the so-called "old evidence problem". The Bayesian element in our account of hypothesis testing is neither necessary nor possible (i.e., consistently applicable), according to him, nor does it very well describe the methods invoked in paradigm cases by the scientists involved to relate specific pieces of evidence to particular hypotheses.[16] Moreover, there is no need, as in our account, for a separate conception of evidence. "Evidence" simply consists of data that, in the way described, confirm or disconfirm hypotheses.

"Glymour Bootstrapping" and Statistical Bootstrapping

Before embarking on an appraisal of his position, we should say how bootstrapping, to be called "Glymour bootstrapping" hereafter, is different from the notion of bootstrapping pioneered by Bradley Efron in statistics[17] since the two are sometimes confused. Glymour's account is called the "bootstrap" account because it exploits the idea of "pulling ourselves up by our bootstraps", i.e., using only resources that individual hypotheses and general theories themselves provide.[18] For his type of bootstrapping, we use a hypothesis of a theory to confirm other hypotheses of the same theory with the help of data, and then use the other hypotheses to confirm the first hypothesis of the same theory. So his account is hypothesis-specific, whereas statistical bootstrapping is data-centric. More explicitly, statistical bootstrapping is a method of generating more data out of the data already collected, to make reliable inferences about a parameter of interest. If the collected data are bad, then obviously one's inference based on bootstrapping those data won't turn out to be good.[19] However, there is no such problem or proviso in Glymour bootstrapping. Since it does not distinguish between "data" and "evidence", if some datum is evidence for a hypothesis relative to a theory, then it will count as evidence for the theory, even though the datum might be bad. Now to focus on statistical bootstrapping. Suppose we have collected a data set from a

---

[16]This is a common complaint, that Bayesian theories have not been employed, even implicitly, to any great degree in the history of science and that therefore they have little to do with actual practices. Our reply, very briefly, is that probabilistic methods were not generally available or very well understood until the 20th century, that traditional methodologies have difficulty in coping with the more recent discovery that a large range of phenomena can only be adequately described in statistical terms, and that many past paradigms of scientific achievement can be more deeply explained if probabilistic concepts are employed to reconstruct them.

[17]See Efron and Tibshirani (1994).

[18]Apparently both "bootstraps" take their names independently from the same American folklore. The folklore has some protagonist doing impossible things by applying his own force to himself. It was the bright idea of Glymour and Efron to extract large philosophical and statistical lessons from turning the entity of interest in on itself.

[19]This issue is not peculiar to statistical bootstrapping. Any statistical inference uses observed data (good or bad) to make inferences about unobserved data.

population. We can generate new "data sets" (called "bootstrap samples") of the same size as the original data set, to mimic repeated sampling of the entire population. This is achieved by re-sampling with replacement from the original data set. The number of bootstrap samples (B) generated is usually 1000 or more, depending on the researcher's choice. Once we have all the bootstrap samples, we can proceed to gather some information about the parameter (i.e., hypothesis) in question. The variability observed in the bootstrap samples is important because this variability mimics the variability that we would see in repeated random samples from the same population (the type of variability we need to construct distributions of test statistics). In statistical bootstrapping, although we make an inference about a parameter (hypothesis), we won't be able to make a reliable inference unless "new data" can somehow be generated from the existing sample. This is why we call statistical bootstrapping data- rather than hypothesis-centric.[20]

Consider an example of Simpson's paradox to show how statistical bootstrapping works. Simpson's Paradox involves the cessation of an association or the reversal of the direction of a comparison when data from several groups are combined to form a single whole. The data in the following table represent acceptance and rejection rates of males and female applicants for graduate school in two departments of an imaginary university in some year.

Table 7.1 shows one formulation of Simpson's paradox where the association between "gender" and "acceptance rates" in the subgroups *ceases* to exist in the combined group. Although the acceptance rates for females are higher in each department, in the combined department's statistics, those rates cease to be different. The data in Table 7.1 are treated as proportions of an infinite population and random samples of size $n = 1800$ are taken from this population. Tables 7.2 and 7.3 show the results of two such random samples from Table 7.1 based on the idea of statistical bootstrapping. They illustrate how a population that exhibits Simpson's paradox can generate random samples that either do or do not exhibit the paradox. This sampling variability and the variability it causes in the estimated proportions is something that must be accounted for while making inferences from a sample to the population. The statistical test procedure must also be able to assess the strength of the evidence for or against H_0 (i.e., there is no Simpson's paradox) based on a single observed sample.

This simple example demonstrates how the bootstrapping technique of generating random samples from a single observed data set works. As we see, those two bootstrapped samples (Tables 7.2 and 7.3) are generated from Table 7.1 as if by pulling oneself up by one's bootstraps, but not by using bits of the same theory to test the other bits of the theory and conversely, as in the case of Glymour's bootstrapping.

---

[20]We owe some insights about statistical bootstrapping to Mark Greenwood.

Table 7.1 Simpson's paradox

Two groups	Department 1		Department 2		Acceptance rates		Overall acceptance rates (%)
	Accept	Reject	Accept	Reject	Depaetment 1 (%)	Department 2 (%)	
Females	90	1410	110	390	6	22	10
Males	20	980	380	2620	2	12	10

Table 7.2 Random sample 1 from population (i.e., Table 7.1) failing to exhibit Simpson's paradox

Two groups	Department 1		Department 2		Acceptance rates		Overall acceptance rates (%)
	Accept	Reject	Accept	Reject	Department 1 (%)	Department 2 (%)	
Females	31	431	42	122	6.7	25.6	11.7
Males	2	284	107	781	0.7	12	9.3

Table 7.3 Random sample 2 from population (i.e., Table 7.1) exhibiting the paradox

Two groups	Department 1		Department 2		Acceptance rates		Overall acceptance rates (%)
	Accept	Reject	Accept	Reject	Department 1 (%)	Department 2 (%)	
Females	24	445	26	118	5.1	18.1	8.2
Males	6	296	111	774	2	12.5	9.9

Confounding

Glymour's account has much to recommend it, and has rightfully loomed large in subsequent discussion of scientific inference and hypothesis testing. Unlike many such accounts, it contains detailed reconstructions of paradigm real-world scientific arguments, from physics to Freud. But it also harbors some difficulties, the ultimate source of which is a failure to distinguish evidence from confirmation in the way that we have.[21]

To begin with, Glymour's "bootstrap account" is not an adequately general account of selective confirmation. It does not deal with statistical hypotheses or techniques and, in particular, with the problem which arises when one hypothesis is mixed up or confounded with another in such a way that the selection of one or the other is problematic. Since the problem is important, perhaps especially in epidemiological studies, an adequately general account of hypothesis testing should deal with it. A rather extended example illustrates both why it is a problem and how our own account accommodates it.

---

[21]For example, Glymour (1980, p. 110): "Confirmation or support is a relation among a body of evidence, a hypothesis, and a theory …: the evidence confirms or disconfirms the hypothesis with respect to the theory."

Although a causal connection between smoking and lung cancer was tentatively accepted after the publication of the Surgeon General's 1964 report, *Smoking and Health*, attempts were made to identify other risk factors, particularly given variance in the distribution of the disease.[22] Perhaps the most prominent of these is that drinking alcohol, particularly beer, is causally linked to lung cancer. Beginning with Porter and McMichael's 1984 hypothesis to this effect,[23] a number of studies indicated that the risk of lung cancer increases as a result of drinking. Data do confirm, that is, raise the probability of, the hypothesis that the risk of lung cancer increases as a result of drinking.[24] The problem is that drinking and smoking are very highly correlated, which is to say that smoking confounds the association between drinking and lung cancer. One way to screen for a confounding effect is to collect data on non-smokers. As soon as the drinking/lung-cancer population is segregated into smoker and non-smoker sub-populations, a distinct correlation (and with it a possible causal connection) between drinking and lung cancer virtually disappears.

To conclude that drinking causes or is a determinative risk factor for lung cancer depends on assuming that it is not mixed up with or confounded with another variable. This is the chief auxiliary hypothesis that must be invoked. But it is confounded by/mixed up with smoking. Therefore, the auxiliary assumption is most likely false. As importantly, selective disconfirmation of it is possible without reaching down for bootstraps. Indeed, it is difficult to see how bootstrapping would help. Showing that the auxiliary assumption in this case is likely false does not depend on appealing to other hypotheses or additional data, but on making a distinction which Glymour does not, between evidence and confirmation. Among non-smokers, the likelihood ratio of developing lung cancer among those who drink as against those who do not is less than 2. This is to say that the evidence for the hypothesis that drinking is a determinative risk factor for lung cancer is very weak compared to the hypothesis that it is not. It should be compared with the likelihood ratio of dying from lung cancer among smokers to dying of lung cancer among non-smokers between the ages of 35 and 84. This ratio is, from one study to the next, 10 or more, very strong evidence for the smoking-lung cancer hypothesis.[25]

The moral of the story is that there are problems of selective confirmation with which the "bootstrap" account does not deal and cannot solve. Glymour is not interested in these problems because they are at stake in few if any of the arguments advanced for generally accepted scientific theories. It is, in Glymour's view, the

---

[22]See, for example, Cochran (1964, pp. 134-55).

[23]Porter and McMichael (1984, pp. 24–42).

[24]See, for example, Chun Chao (2007), for more detail and references.

[25]Bagnaardi et al. (2011).

task of confirmation theory to make both perspicuous and persuasive the inferential structure of such arguments. That our own account of hypothesis testing can do so in the case of observational studies is a mark in its favor.

Circularity

Duhem explicitly exempted observational studies from his thesis that confirmation always required auxiliaries and so too, at least implicitly, does Glymour.[26] Their focus is on the testing of theoretical hypotheses, and on the need for auxiliary hypotheses to derive observational implications from them. In the case of Glymour, the need for "bootstraps" is more specifically focused on the need to determine values for constants which theoretical hypotheses often contain and which do not submit to direct measurement. As we have seen, he proposes a way to do so that does not require recourse to "auxiliaries" drawn from other theoretical contexts. But even in the case of testing theoretical hypotheses, in particular of determining the value of the theoretical constants in them, there are difficulties with the "bootstrapping" account.

The most significant of these difficulties would seem to be that the "bootstrapping" account is circular, or, if not, would seem to avoid circularity only by way of appealing, against Glymour's stated intention, to probabilities.

We can borrow another of his examples to make the point clear. To test the correctness of the law of universal gravitation, $F = Gm_1m_2/R^2$, we need to determine F (force). To do so we can use the second law of motion, $F = ma$. We can measure the mass and the acceleration of a body in motion, and using Cavendish's estimate of the constant of gravitation, G, determine whether the values for F in the two laws are congruent. The difficulty is that it is possible that the two values are congruent but that each of the laws is false; it might just so happen that the measured values of the other variables together with the constant produce the same number in the case of the experiments carried out, but that this is simply a coincidence or the result of self-canceling errors. As Glymour himself points out, in the 17th century confirmations of Kepler's First Law of Planetary Motion required assuming the correctness of the Second Law, and confirmation of the Second Law required assuming the First Law. In fact, many astronomers were unclear whether errors in one were compensated by errors in the other. It was not until the invention of the micrometer and Flamsteed's observations of Jupiter and its satellites that the Second Law could be confirmed (in some general sense of the word) without using the First Law. Of course, the probability of the coincidental or mutually-compensated congruence of the two

---

[26]In *Theory and Evidence*, Glymour assumed (in part for the sake of argument) that Hempel's "positive instance" account of confirmation correctly characterized the evidential relation between data and hypothesis in observational studies and that bootstraps were necessary only when the hypotheses tested contained theoretical terms whose values could not be directly measured but were inferred.

inferred values for force is very small, and becomes smaller as a function of the number and variety of the experiments. But to say this is to admit that an adequate account of testing must include probabilities.[27]

A Worry not Dispelled

Glymour acknowledges the circularity, but does not think that it is vicious. It would be vicious if "*whatever the evidence* the hypothesis would be confirmed…".[28] But as we have seen, bootstrap testing is so characterized as to require the possibility of evidence that tells against as well as for the target hypothesis. There is nothing in any of the arguments that Newton used to derive testable instances of the law of universal gravitation which guaranteed its success in advance. The instances had first to be derived and then matched with observation. "Still", Glymour continues,

> "one is troubled by the following kind of worry. Suppose that … *A* is used together with evidence *E* to justify *B*; and suppose that … *B* is used with evidence *E'* to justify … *A*; then might it not be the case that *E* and *E'* determine neither a real instance of *A* nor a real instance of *B* but provide instead *spurious* instances, which appear to be correct because the errors in *A* are compensated for in these cases by errors in *B* and vice versa? Indeed, it might be the case, but it would be wrong to infer straight out from this that *E* and *E'* provide no grounds at all for *A* and *B*.[29]

This makes surface sense, but from our perspective there are two underlying confusions. One is between "real" and "spurious" instances. An *instance* is characterized in terms of its syntactical form and deductive relationship to a hypothesis, quite independently of whether it is "real" or not. The addition of the adjectives seems to be driven by a "true-model" assumption, that a "real" instance would, if verified, give us good reason to believe that the hypothesis is true, an implication further reinforced by the word "justify". The other confusion is that while *E* and *E'* in the case imagined *seem* to confirm *A* and *B*, they do not really do so because (Glymour implies) at least one of the premises of each inference to an instance was false. But the intuition that they provide some "grounds" can be preserved once we recognize that *E* and *E'* constitute *evidence* for the hypotheses tested (against alternatives) even though the hypotheses might be false. As we have said more than once, evidence unlike confirmation does not incorporate a "true-model" assumption. It is only when the two concepts are mixed together that a retreat to language which unsuccessfully tries to steer a course between both is forced. When at the end of the discussion, Glymour says that certain philosophers were afraid of circularity

---

[27]"Perhaps we should infer instead that there are better and worse justifications, and that arguments or confirmations (or, as I prefer, tests) which guard against compensating errors provide better grounds than those which do not" (Glymour 1980, p. 108).

[28]*Ibid.*

[29]*Ibid.*

"because of a commitment to a particular and unnecessary conception of justifi-cation," he could have added that part of this commitment involved the idea that data could provide evidence for a hypothesis only if they also confirmed it.

Variety of Evidence

Glymour tries to turn the appearance of circularity to his benefit by arguing that it underlines the necessity of many and varied tests for each of the hypotheses employed to determine measurable values for the theoretical constants in the instances that confirm or falsify the hypotheses we eventually accept. The greater the number of and the more varied such tests, the better,[30] among other reasons because they then serve to separate hypotheses. As he notes, with the micrometer and Flamsteed's observations of Jupiter and its satellites, confirmation of Kepler's second law could at last be obtained without any assumptions concerning the planet's orbit. Again we agree with him "that the development of any complex modern theory can[not] be understood without some appreciation of the way a variety of evidence serves to separate hypotheses".[31] We have stressed the point. Whereas it is for Glymour a way of mitigating the possibility of off-setting errors in hypotheses both used to derive potentially confirming instances of a theory, it is for us definitive of what counts as evidence. Two corollaries follow immediately. One is that is that it makes little sense from our point of view to say, as he does, that "part of what makes one piece of evidence relevantly different from another piece of evidence is that some test is possible from the first that is not possible from the second…" (Ibid., p. 140). But this does not cohere very well with the view just stated that a variety of evidence serves to separate hypotheses. Moreover, it is clear that Glymour is running "evidence" and "data" together here, the inevitable result of conflating evidence with confirmation; "pieces of evidence" cannot be identified apart from particular testing contexts, although data can. *Data* are relevantly dif-ferent from one another in that they can be used to provide different tests of the same hypothesis, and if positive increase the degree of its confirmation. On the positive instance account that Glymour favors, the more such data the better. But *evidence,* which is the means whereby we contrast hypotheses with one another, can (in context) itself be weighed. We say this is "better evidence" than we had before, and mean that the data are more likely on the hypothesis than on its relevant alternatives. "Better data," on the other hand, means that the data are less con-taminated, less subject to bias, and so on.

---

[30]Once again, Glymour would seem to depart from his purely qualitative account in underlining an ordinal intuition.

[31]*Ibid.*, p. 141.

The other corollary of the fact that for us, but not for Glymour, separation of hypotheses is definitive of evidence also has to do with its variety. "If," he has argued,

> a given piece of evidence may be evidence for some hypotheses in a theory even while it is irrelevant to other hypotheses in that theory, then we surely want our pieces of evidence to be various enough to provide tests of as many different hypotheses as possible, regardless of what, in historical context, the competing theories may be" (*Ibid.*).

We have already argued that "pieces of evidence," however they might be individuated on the "bootstrap" account, are not context-free. Data constitute evidence for/against a hypothesis (in context) only to the degree that they "fit" the hypothesis; it is possible for data to be irrelevant, not for evidence. But more to the immediate point is that "variety" here is understood in terms of the confirmatory isolated tests of individual hypotheses, the more the better. Whereas on our view, "variety" is understood, more plausibly we believe, as a function of what allows us to weigh and measure competing hypotheses with respect to the available data. That Kepler did not simply determine elliptical orbits for the planets as the best fit for the data but gave a physical argument for the area rule, his second law, has to do with its confirmation. That the orbits so determined were ellipses provided evidence for Newton's theory against its contemporary alternatives. Glymour's claim that we provide evidence for hypotheses regardless of what the competing theories may be is mistaken, we believe, as a matter of history, but also as a guide to how scientists should proceed.

The Importance of Determining the Value of Theoretical Constants

Lest there be any doubt about it, we agree with Glymour that the determination of theoretical constants and the obtaining of congruent values for them as a result of testing a variety of different hypotheses is an indispensable component of theory acceptance. Although he does not mention it in *Theory and Evidence*, Jean Perrin's obtaining a consistent value for what he termed Avogadro's Constant in a number of different hypotheses as a result of some very sophisticated experiments, among them Einstein's regarding Brownian motion, was the single most significant indication of the empirical success of the atomic theory of matter.[32]

"Selectivity" as Glymour understands it has a role to play: it allows for the testing of hypotheses that contain theoretical constants. So too does "severity" when it is possible to assess the statistical adequacy of individual models. But in our view the *separability* of hypotheses that evidence makes possible is much more both general and fundamental. One way to bring this out in the case of Glymour is to note that he,

---

[32]See Nye (1972).

like Duhem, thinks the main problem in theory testing is to "bridge the gap" between observational generalizations and theoretical laws. "Bootstraps" are designed to limit the resort to bridging "auxiliaries," and thus to calm philosophical worries that such auxiliaries might eventually include all of science. But as we pointed out in this chapter, the "auxiliaries' problem" arises equally at the level of observational hypotheses and impacts the day-to-day practice of science generally. In our paradigm case, it is how to distinguish between the alcohol-lung cancer and smoking-lung-cancer hypotheses. In an earlier case, it was how to separate the anthropogenic and natural-rhythm global warming hypotheses. In our view, the main (although certainly not the only) aim of experimentation is to discriminate between hypotheses. This is the necessary first step in testing them. Jupiter's moons discriminated between the Copernican and Ptolemaic hypotheses in a way that no previous experimental data were able to. Darwin's problem was similarly not to find data which confirmed the theory of Natural Selection, but to find data which allowed other biologists to distinguish between his theory and Lamarck's.[33] The big and even more difficult problem was to find data that could discriminate between the General Theory of Relativity and Newton's gravitational theory. Glymour has part of this right when he focuses on "selectivity," but he seizes on the wrong reasons ("irrelevant conjunctions" and other philosophical conundrums), misleadingly limits his account to theoretical constants, and has no principled way in which to separate hypotheses, still less to quantify the way in which data provide evidence for and against them without at the same time being called upon to confirm them.

References

Bagnardi, V., Rota, M., Scotti, L., Jenab, M., Bellocco, R., Tramacere, I., et al. (2011). Alcohol consumption and lung cancer risk in never smokers: a meta-analysis. *Annals of Oncology, 12,* 2631–2639.

Chao, C. (2007). Associations between beer, wine, and liquous consumption and lung cancer risk: a meta-analysis. *Cancer and Epidemiological Biomarkers, 16,* 2436–2447.

Cochran, W. (1964). The planning of observational studies of human populations. *Journal of the Royal Statistical Society, Series A, 128,* 134–155.

Douven, I., & Meijs, W. (2006). Bootstrap confirmation made quantitative. *Synthèse, 149*(1), 97–132.

Duhem, P. (1954). *The Aim and Structure of Physical Theory (translation of the second edition of the French original published in 1914).* Princeton: Princeton University Press.

Efron, B., & Tibshirani, R. (1994). *An Introduction to the Bootstrap.* New York: Chapman and Hall/CRC Monographs on Statistics and Applied Probability.

Glymour, C. (1980). *Theory and Evidence.* Princeton: Princeton University Press.

Glymour, C. (2010) In Mayo & Spanos (Eds.), *Error and Inference* Cambridge: University Press Cambridge.

Kuhn, T. (1962/1970/1996). *The Structure of Scientific Revolutions.* Chicago: University of Chicago Press.

---

[33]See Sarkar (2007) on these and related issues.

Laymon, R. (1983). Newton's demonstration of universal gravitation and philosophical theories of confirmation. In Earman, J. (Ed.), *Testing Scientific Theories,* Volume X in *Minnesota Studies in the Philosophy of Science* (Minneapolis: University of Minnesota Press).

Nye, M. (1972). *Molecular Reality*. Amsterdam: Elsevier.

Porter, J., & McMichael, A. (1984). Alcohol, beer, and lung cancer—a meaningful relationship? *International Journal of Epidemiology, 13*, 24–42.

Sarkar, S. (2007). *Doubting Darwin? Creationist Design on Evolution*. New York: Wiley-Blackwell.

Chapter 8
Veridical and Misleading Evidence

Abstract Like the error-statisticians, Glymour, and us, Peter Achinstein rejects an account of evidence in traditional Bayesian terms. Like the error-statisticians and Glymour, but unlike us, his own account of evidence incorporates what we have called the "true-model" assumption, that there is a conceptual connection between the existence of evidence for a hypothesis and having a good reason to believe that the hypothesis is true. In this connection, and unlike any of the other views surveyed, Achinstein does not so much analyze the concept of evidence per se as provide a taxonomy of conceptions of evidence—subjective, objective, potential, "epistemic-situational," and "veridical." He then both argues that only "veridical evidence" captures exemplary scientific practice and explains why this should be the case. In the first half of this chapter, we set out Achinstein's criticisms of the Bayesian account, then develop and in turn criticize his own account of "veridical evidence." In the second half of the chapter, we contrast "veridical" with "misleading" evidence and show how two key theorems concerning the probability of misleading evidence allow us to measure it.

Keywords Veridical evidence · Misleading evidence · "Good reasons for believing" · Accumulation of evidence · Probability of misleading evidence

The Bayesian Account of Evidence Rejected

In Achinstein's view, developed over time in (1983), (2001), and (2005), but in its essential features unchanged, the traditional Bayesian account of evidence, that D is evidence for and thus confirms H just in case $\Pr(H \mid D) > \Pr(H)$, is both too weak and in a certain sense too strong. It is too weak since counter-examples are easily generated which show the intuitive gap between providing evidence for and raising the posterior probability of a hypothesis. Typical of the counter-examples is that a passenger increases the probability that she will die in a crash when she boards an

© The Author(s) 2016
P.S. Bandyopadhyay et al., *Belief, Evidence, and Uncertainty*,
Philosophy of Science, DOI 10.1007/978-3-319-27772-1_8

airplane, but boarding an airplane is not evidence that she will die in this way. Similar, albeit somewhat more complicated, examples show that D can provide evidence for H without thereby raising the probability of H[1] Indeed, on occasion they can lower it.

But the Bayesian conception of evidence is also in a sense too strong, for it incorporates the idea that what counts as evidence is relative to the subjective beliefs of particular individuals before and after data have been gathered. As such, it runs afoul of what Achinstein styles *A Principle of Reasonable Belief*, to wit, that "If in the light of background information b, e is evidence that h, then, given b, e is at least some good reason for believing h." In a Kantian vein, if *e* is a good reason for an agent *a* to believe that *h*, it must be equally a good reason for everyone else to believe that *h* whether or not it is the reason why *a* believes that *h*. It is only in this way that the objectivity of evidence, and hence of the claims which it is invoked to support, can be safeguarded.

Veridical Evidence

As his description of it suggests, Achinstein requires that evidence is "veridical" only if it is true and the hypothesis for which it is evidence is true as well. As we have already seen, a third requirement, in his view necessitating the first two, is that "e provides a good reason to believe that h" (Achinstein 2001, p. 24). In turn, "if e is a good reason to believe h, then e cannot be a good reason to believe the negation of h" (Achinstein 2001, p. 116), which comes to saying that "e is a good reason to believe h only if Pr(h | e) > ½" (Achinstein 2001, p. 150).[2] But these conditions for veridical evidence, though necessary, are not also sufficient. "Let e be the information that a man eats Wheaties for breakfast and h the hypothesis that this man will not become pregnant. The probability of h given e is extremely high...But e is not evidence for h." (Achinstein 1983, p. 154). To claim otherwise would be to make a bad joke. To bar this sort of counter-example to his first three requirements, Achinstein adds a fourth, that there is an "explanatory connection" between the truth of h and e.

Before criticizing this conception of evidence, two brief comments should be made about the resort to the notion of an "explanatory connection." Although Achinstein provides various definitions of it, all are problematic, as is the notion itself. Indeed, for Hume and many of his positivist followers, an "explanatory

---

[1]Achinstein does not make our distinction between data and evidence. In what follows, we have collapsed the distinction between them and used his "*e*" to stand for both. It is, of course, our position that a distinction between them should be observed.

[2]Achinstein is working here with the notion of an objective epistemic probability, independent of what people happen to believe. Our criticisms of his position, and the examples on which they sometimes depend, do not turn on questions concerning the correct interpretation of the probability operator. See Achinstein (2010) on this topic where he discusses Mayo's work.

connection" is itself no more than probabilistic; to say that A (an event, say) explains B (another event) is to say that A raises the probability, hence the predictability, of B. Of course, this is not the end of the matter; the positivist gloss has itself been much criticized. But at the very least it is fair to say that the notion of an "explanatory connection" is not more basic than the notion of evidence which it is intended to clarify. The other comment is that our own account of evidence disposes of the counter-example and others like it without further ado. The likelihood of eating Wheaties on the hypothesis that a man is pregnant is neither greater nor less than the likelihood of eating Wheaties on the hypothesis that a man is not pregnant, and therefore eating Wheaties does not constitute evidence for either hypothesis. Further analysis of such otherwise murky notions as "explanatory connection" is not necessary.

In any case, Achinstein thinks that scientists seek veridical evidence in his sense because they seek the truth, and they want good reasons for believing that particular hypotheses are true. Veridical evidence as characterized both guarantees truth and provides such good reasons. This is a particularly clear example of what we earlier referred to as the "true model" assumption. It is at the root of the criticisms to follow.

Good Reasons for Believing

The main difference between potential and veridical (read: "genuine") evidence is that the second but not the first requires the truth of the hypothesis for which it is evidence. Only in this way can it provide good reasons for believing that the hypothesis is true. But the insistence on truth is problematic.[3] Take Newton's derivation of Kepler's laws from his gravitational theory as a familiar example. Kepler's Laws are not, strictly speaking, true, that is, they furnish us with slightly idealized models which rather closely approximate the observed motions of the planets around the sun. Nor is Newton's gravitational theory "true" in any usual sense of the word. Observed deviations from it are small when the ratios of the gravitational potential and of the velocity of planets to the speed of light are small; when the ratios are large, Einstein's general theory of relativity must be invoked. The derivation of Kepler's laws goes through only if a number of assumptions are made—that the Sun and the planets can be treated as uniformly spherical point-masses, that the net gravitational force on a planet is toward the Sun (in which case the mathematically-intractable attraction of other planets may be ignored), that electromagnetic and other forces acting on planets do not need to be considered—which are in fact false. And so on. Yet the derivation of Kepler's Laws was taken by Newton and his followers as the principal piece of evidence in behalf of the

---

[3]While we take issue with some of its main themes, Cartwright (1983) has a detailed and very influential discussion of the respects in which physical laws are not true, and the implications thereof.

gravitational theory. The analysis of concepts in the philosophy of science is both descriptive and normative;[4] we cannot very plausibly contend that neither Newton nor succeeding generations of scientists for whom his "proof" was taken as an evidential standard of empirical success were misguided. It won't do to say that the derivation *was* evidence and is so no longer (for example in the light of the general theory of relativity), for Achinstein rightly maintains that evidence is agent- (and thus era-) independent; evidence is evidence.[5] Nor does it help to weaken the truth requirement, as Achinstein does on occasion (e.g., Achinstein 1983, p. 169), to probability-of-truth, for it isn't that either Kepler's or Newton's Laws are "probably true," but that they are idealized approximations (and therefore "false") of what we

---

[4]In her highly readable and very influential book, *Science as Social Knowledge*, Helen Longino disagrees with our dual emphasis on the normative and descriptive aspects of philosophy of science. She addresses descriptively the evidential issues in which she is interested in a "non-philosophical discussion of evidence." She chides philosophers for their use of formal and normative models "*that are never realized in practice*" (Longino 1990, emphasis ours). We differ from her in at least two ways. First, we think that without formal and normative accounts, it is difficult to assess a variety of scientific and philosophical issues. It is only against the background of such an account, for example, that we have been able to contend that across the board application of the collapsibility principle leads to the apparent paradoxical nature of Simpson's Paradox (Bandyopadhyay et al. 2011). Second, there are many, many counter-examples to her claim, among which the enormously influential work of Lewis and Sally Binford on the theory and practice of archeology. See Binford and Binford (1968). In the Appendix to the last chapter of this monograph, one of the authors of the monograph, a working ecologist, shows how our formal/normative accounts of confirmation and evidence play a role in current ecological science and beyond.

[5]Longino takes issue with the claim, which we share with Achinstein, that evidence is independent of any agent's beliefs and thus objective. According to her, all evidential relations are "*always determined by background assumptions*" (*ibid.,* p. 60, emphasis ours) and therefore the same data can provide different evidential support for two competing theories simply in virtue of the fact that they make different background assumptions. This leads her to brand her account as "a contextualist analysis of evidence." Three points are worth mentioning. First, our account is also contextualist, but in a very different and more minimal way: data provide evidence for one hypothesis (and its auxiliaries) against a rival hypothesis (and its auxiliaries) if and only if the likelihood of the data on one hypothesis as against the other is greater than one. Our account of evidence is local, precise, and comparative. Her account is general, rather open-ended as to what counts as a "background assumption," and (at least not explicitly) non-comparative. There is a clear sense in which our account, unlike hers, is objective and agent-independent. Second, even though many of the auxiliary assumptions (a.k.a. background assumptions) of Newtonian and Einsteinian theory differ, the derivation of Mercury's observed perihelion shift is nonetheless considered evidence for the latter as against the former (see our discussion of the "old evidence" problem in Chap. 9). If we were to take her account seriously, a serious shift in scientific practice and the assessment of the evidential support for theories would result. Third, in company with most other philosophical analyses of them, she tends to identify the concepts of "evidence" and "confirmation." After subjecting the Hempelian account of confirmation to severe criticisms for its syntactic character, she goes on to write that the right question to ask is: "[c]an this definition of the confirmation relation be the source of the relation between evidence and hypothesis?" (*ibid.* p. 24). But as the tuberculosis example makes clear, the definition of "confirmation" is not the source of the (properly understood) relation between "evidence" and hypothesis; "confirmation" and "evidence" can vary dramatically.

in fact observe (itself idealized and formatted in certain ways). Indeed, it is common practice for physicists (and perhaps even more common for scientists working in disciplines where universal as against statistical generalizations are rare) to use a number of different mathematical models to describe the same phenomena as a function of the particular questions they want answered, even when the models are incompatible and hence cannot all be "true." What is crucial is not that the models are true, but that they can be tested, perhaps most often by deriving predictions from them and, as we mentioned earlier, the values of the parameters in them estimated. On our view, such testing is where the concept of evidence takes hold, and always involves a comparison between models.

Achinstein links the truth requirement to having good reasons to believe a theory and both of them to evidence. But we have argued that evidence moves in a different direction. This can be brought out by a brief re-consideration of the tuberculosis-testing paradigm. Chest X-rays are still the standard test for the presence of tuberculosis in the lungs. A positive test indicates the presence of TB, which is to say that it is taken as evidence that the person who tests positive has tuberculosis. As we pointed out earlier, the likelihood of testing positive on the hypothesis that TB is present is 26 times greater than the likelihood of testing positive on the hypothesis that it is not. Yet as we also pointed out, the posterior probability of a positive test result on the hypothesis that TB is present is low, well below the 0.5 threshold that Achinstein posits. There is a conflict of intuitions here.[6] On the one hand, a positive result from a chest X-ray is generally taken as evidence of the presence of TB, and thus on Achinstein's account should provide a good reason for believing that an individual is infected (and appropriate further measures taken). But the posterior probability of being infected given a positive result is nonetheless low, in which case and again on his account, the individual tested does not have a good reason to believe that she is infected. Rather than arbitrarily say that on the basis of armchair reflections there is no evidence of TB and therefore there is no good reason to believe that there is, we think it much better to separate the having of good reasons from the question of what constitutes evidence. If one were then to ask, "but why is evidence desirable if not to provide such reasons?" we would answer, to measure the comparative weight of the data gathered in determining which one of a pair of hypotheses is better supported by it. Weighing and measuring the data is a crucial part of scientific testing, quite independent of providing good reasons to believe that the hypotheses being tested are true or false.

---

[6]Suppose what could be the case, that the test has been improved so that its sensitivity and specificity are now both 0.999. $\Pr(H \mid D \ \& \ B)$ is now 0.085, whereas the $\Pr(H)$ is as before 0.000093. The posterior probability is very low, and yet the LR is 999. Confirmation and evidence pull us in very different directions. In fact, the probability of misleading evidence this strong is only 0.001 (Taper and Lele 2011). At this point, a doctor and patient might wonder whether we need to know more about the patient. Has she recently been abroad? If so, then it might not be wise to identify her with the general population in which the prior probability of TB for an individual randomly selected is 0.000093. A wedge between confirmation and evidence is easily driven.

Good reasons to believe have to do with beliefs, however generalized they might be. Evidence does not. In conflating them, Achinstein would have us say of particular cases that we both have and do not have good reasons to believe that a hypothesis is true on the basis of the same data.

It should help to illustrate our differences in the above respect if we focus more specifically albeit briefly on the example Achinstein takes as paradigm in (Achinstein 1983). Alan's skin has yellowed, on the basis of which the examining doctor tells him on Monday that he has jaundice. Additional tests are run, as a result of which the doctor says four days later that Alan does not have jaundice, but that his yellow skin was caused by a yellow dye with which he had been working. Achinstein proceeds to ask three questions concerning what the doctor should affirm on Friday:

1. Alan's yellow skin was evidence of jaundice and still is.
2. Alan's yellow skin was but no longer is evidence of jaundice.
3. Alan's yellow skin is not and never was evidence of jaundice.

As we might expect, his answer is 3; 1 is ruled out since the hypothesis is false and does not explain why Alan's skin yellowed and 2 is true of the doctor (it was evidence for him on Monday but not Friday) but not generally or objectively. But this answer is seriously counter-intuitive. Yellow skin is evidence of jaundice in the precise sense that it warrants further tests for the underlying increased concentration of bilirubin in the blood. If the blood workup came back negative for bilirubin we would not say that the yellow skin was not evidence after all, but that it was misleading evidence. In courts of law, circumstantial evidence that the accused did it which is eventually trumped by the discovery of a more likely suspect is nonetheless evidence, although far from conclusive, that the accused did it. There are two ways in which to shore up this intuition and make it more precise. Both have to do with the mistaken character of the questions asked. First, the question to be asked is not whether data constitute evidence, potential or veridical, but whether they constitute evidence to what degree. Evidence is a qualitative concept, but in specific cases its strength can also be measured. It is by the same token a comparative concept. Yellow skin is evidence for the hypothesis that Alan has jaundice as against the hypothesis that he does not. But it is not evidence, other things being equal, that he has jaundice as against the hypothesis that he has come into contact with yellow dye. In any sort of plausible scenario, the doctor would begin by going through available alternative hypotheses and ask Alan, among other things, whether he had been working with dyes recently. Then he would order additional tests to discriminate or what we called in Chap. 6, "separate" between these hypotheses, that is, to evaluate the strength of the evidence for each. Even then the doctor might misdiagnose the case. But a misdiagnosis does not entail, as Achinstein would have it, that what the doctor thought was evidence in fact was not. It was just inconclusive evidence.

Quantitative Thresholds, the Accumulation of Evidence, and "Good Reasons to Believe"

As we noted, Achinstein's account of evidence has a quantitative, as well as a truth and explanation, component, that the probability of the hypothesis given the data is >0.5. In the section "Absolute and Incremental Confirmation" in Chap. 2, we set out our objections to any account of evidence in terms of thresholds of the probability of hypotheses on the data. Among other things, it mixes together questions concerning evidence and confirmation, does not allow for the comparison of the evidential strength of data relative to different hypotheses, and, perhaps most importantly, rules out the possibility that the posterior probabilities of incompatible hypotheses can meet the threshold when more than two hypotheses are at stake. But our criticism here is that in disallowing evidence that fails to meet the threshold as even potential evidence, it bars the door to the accumulation of such evidence in a way that eventually provides us, in the Achinsteinian terms that we reject, with "good reasons" to believe a hypothesis.

Suppose a coin is tossed five times and comes up with heads four times. We consider two models: a "biased coin model" with the probability of heads equal to 0.8, and a "fair coin model" with the probability of heads equal to 0.5. The sequence of four heads and one tail is intuitively evidence for the biased coin model relative to the fair coin. However, and again intuitively, it is weak evidence because there is a reasonable chance that a fair coin might also give such a sequence. But it should not for that reason be discounted as potential evidence. As we keep insisting, whether data constitute potential evidence depends on context just as does their relative strength, a point that tells against qualitative accounts generally.

One reason why it is important to consider weak evidence is that while a single small piece of evidence by itself need not be compelling, many small pieces of evidence could be accumulated so as to result in strong evidence. Consider a scenario in which four people drawing at random from an urn which may contain either biased (probability of heads = 0.8 as in the previous example) or fair coins. Their first task is to flip each coin five times after drawing each of them at random from that urn, and then to infer from those data whether the urn contains an equal number of unbiased and biased coins. Since none of them knows which coin is biased or fair when each draws coins from the urn, the appropriate prior probability for each coin to be biased or fair on each draw should be 0.5 Since each draw of a coin as fair has been assumed to have a prior probability of 0.5, it follows that the prior probability of that urn having an equal number of fair and biased coins should also be 0.5. Suppose that each draws five coins at random and when they flip each of them one time, all end up with the same result: four heads out of five flips. The evidence for a biased coin in each person's sequence of tosses is weak from the perspective of our account of evidence since the biased/fair LR = 2.62. However, the combination of their evidence is very strong, i.e., LR (biased/fair) = 123.4.[7] Moreover, when the data are pooled, each person's degree of confirmation that the urn contains an equal number of biased and fair coins has changed from a prior probability of 0.5 to an a posteriori probability of 0.9919, which provides a very good reason to believe that the urn is filled with more biased coins than fair ones. As many pieces of weak evidence combine to produce strong evidence against a hypothesis, different agents' degrees of belief that a hypothesis is weak could also be up-dated to produce a strong confirmation amounting to "a good reason to believe" that the hypothesis is false.

In the preceding paragraph, we touched on the concepts of accumulation of evidence and updating an agent's degree of belief. The accumulation of evidence is a powerful tool, to be distinguished from the idea that an agent updates her degree of belief in light of new data. In an up-dating account of confirmation, the agent's degree of belief change, as already stated, must conform to the principle of conditionalization which relies, among many other factors, on her subjective prior distribution of degrees of belief over alternative hypotheses. In contrast, the conception of accumulation captured in our account of evidence does not depend on any agent's prior distribution of degrees of belief over alternative hypotheses or in the certainty of the data themselves. Objectivity is preserved.

Achinstein emphasizes "veridicality" because he wants evidence for a hypothesis to furnish good reasons for believing that the hypothesis is true. But in requiring as part of his account that the hypothesis be true, he departs rather sharply from the exemplary scientific practice he wants to describe. Moreover, in not allowing as "potential" evidence data which fail to meet a threshold requirement, he rules out the way in which bits of evidence are accumulated to make the case for particular hypotheses. Finally, from our point of view he misses the fundamentally comparative and contextual character of evidence, the basis of the virtually-universal intuition that some evidence is better than others. But the "true-model" assumption that guides his and some other accounts of evidence is also incapable of coping with what might be called "misleading evidence."

Misleading Evidence

Achinstein insists that genuine evidence has to be error-proof in this sense, that if *e* is to be evidence for *h*, then *h* as well as *e* must be true. His main reason for doing so is that otherwise evidence could not provide good reasons for believing that *h* is true. Our account of evidence is weaker in the same sense, that *e* can be evidence for *h* even though none of them—*e, h,* or its alternative *h'*—is true. That is, the data

---

[7]Notice this evidence is the same regardless if one computes the LR for the two models on the basis of 25 throws, or as the product of the LR's for the five sequences.

can constitute evidence for a false hypothesis. We can call such evidence "misleading."[8] Achinstein rules out its possibility in principle, but in so doing he turns his back on legal and scientific practice, not to mention ordinary language, in which what is construed as evidence has been assembled for a great number of claims and hypotheses which turn out to be false (however much they seemed to be true at the time). In almost all of these cases (to generalize rather blithely), the evidence assembled was not the result of misperception or fraud;[9] rather, it was eventually supplanted by better, stronger evidence for competing claims and hypotheses. Given what we take to be this generally-shared supposition, that in context even "strong evidence" can be misleading, it is necessary to ask whether a more precise account of misleading evidence can be given. We think that it can, and that any adequate epistemology must do so (just as, we argued in Chap. 2, that any adequate account of scientific evidence must include a theory of error).

But before providing our own probabilistic account of misleading evidence,[10] we want to make three preliminary comments.

First, although the notion of truth plays an indispensable role in a broad array of human activities, it is of limited use in contemporary mainstream science. In this respect (although not in all others) we follow Popper, for whom the most we can say about scientific hypotheses is that some have held up better than others when subjected to repeated and, ideally, rigorous tests. Once burned, twice afraid. One hypothesis after another in the history of science has been taken to be "true," only to be rejected in the light of new data or alternative hypotheses which better accounted for the old. Much the same sort of thing can be said about the helpfulness, or lack thereof, of the concept of "knowledge" in science. It is not simply that we are fallible creatures, but that indeterminacy of various kinds is built into the fabric of the universe.

Second, there is no point in denying that the assertion that snow is white is true or that I know that the earth is (virtually) round. But scientific hypotheses and the data enlisted in their support are much more complex. As we have already argued, all universal generalizations necessarily outrun the finite evidence for them and are therefore always potentially false, those scientific generalizations which are taken as *laws* and customarily invoked to explain the phenomena are idealizations and approximations and therefore not strictly true, and data are inevitably subject to

---

[8]See Royall (1997, 2004): "The positive result on our diagnostic test, with a likelihood ratio (LR) of 0.94/).02 = 47, constitutes strong evidence that the subject has the disease. This interpretation of the test result is correct, regardless of that subject's actual disease status. If she does not have the disease, then the evidence is misleading. We have not made an error—we have interpreted the evidence correctly. *It is the evidence itself that is misleading*".

[9]A famous case of the former involved the French physicist Pierre Blondlot's continued perception of "N-rays" in his experimental set-up even after the American physicist Robert Wood had surreptitiously removed parts of the apparatus necessary to see them. See Nye (1980). Cases of fraud which occasionally come to light involve the deliberate fabrication of data.

[10]One of the few philosophers to take misleading evidence seriously is Earl Conee (2004), although his account (and what he styles an "evidentialist" approach to epistemological issues) very much differs from our own.

error and must be corrected by extensive statistical analysis before they can be of any real use. It would be unhelpful to say that all scientific hypotheses and complex data sets against which they are tested are "false," and therefore that all scientific evidence is misleading. For one thing, it would wrongly open the door to those who claim that scientific hypotheses such as the theory of evolution or quantum theory have very little credibility. Our point is that though some scientific evidence is "misleading," this fact does not preclude our being able to weigh and measure it in such a way that some hypotheses are (much) better supported, their parameters more exactly determined, than others. In Chap. 2, we characterized evidence as a data-based estimate of the relative divergence of two models from the truth. On this characterization, evidence is misleading if it mistakes which model is closer to the truth.

Third, given what has been just said, an adequate account of misleading evidence would quite naturally be probabilistic in character. Not only does probability theory provide the tools to shape, sharpen, and reformulate an important epistemic concept, it also allows the derivation of theorems which bring the apparent waywardness of misleading evidence under some formal control.

The Probability of Misleading Evidence

On a probabilistic account, it is always possible that the evidence for a hypothesis is misleading in the twofold sense that the evidence, no matter how strong, is consistent with the falsity of the hypothesis, and that the data constituting the evidence are more probable on a competing hypothesis than on the hypothesis most similar to the mechanism actually generating the data. Once again, consider the tuberculosis example from Table 8.1.

The datum, that a particular individual has a positive result from a chest X-ray is strong evidence on our account that the subject has TB, i.e., the ratio of the likelihoods of the datum on the hypothesis that she does as against the hypothesis that she doesn't = 0.7333/0.0285 ≈ 26, strong evidence indeed. As we noted earlier, this data-based inference is correct regardless of whether she really has the disease. However, the observation of the positive test result could itself be misleading regardless of the fact that it provides strong evidence for the hypothesis. If the disease is absent, the Table 8.1 shows that the probability of a misleading positive test is only 0.0285, i.e., the probability of a positive test result when the disease is in fact absent is very low.

Table 8.1 A summary of TB X-ray test results

	Positive	Negative
Disease present	0.7333	0.2667
Disease is absent	0.0285	0.9715

Table 8.2 A summary of fair/biased coin flip results

	Positive	Negative
The coin is biased	0.4096	0.5904
The coin is unbiased	0.1562	0.8438

This point can be reinforced in connection with a slightly more developed example. Consider two hypotheses, "the fair coin" hypothesis, that the probability of heads equals 0.5, and "the biased coin" hypothesis, that the probability of heads equals 0.8, to see how the account incorporates accumulated evidence which could at the same time be misleading. The first sequence of coin tosses with four heads and a tail is taken to be evidence for the biased coin hypothesis over the fair coin hypothesis. The LR of the biased/fair = 0.4096/0.1562 = 2.62 provides evidence for the biased coin hypothesis, but it is weak. If we take positive test results in tossing a coin to provide evidence that the coin is biased and negative test results to provide evidence that the coin is unbiased, then we get Table 8.2.

However, if the biased coin hypothesis is false, then the probability of misleading evidence is 0.1562. After receiving 4 heads out 5 tosses, suppose we have received the following sequences of heads in each of 5 tosses by five people; 2, 3, 2, and 1 heads yielding 12 heads out of 25 tosses. After we have gathered more evidence, we have now received very strong evidence for the fair coin hypothesis over its biased coin alternative. The LR (Fair/Biased) = 0.155/0.000298 = 529.396 times. In the first sequence of 4 heads out of 5 heads, there is weak evidence for the biased coin hypothesis. However, as we gather more evidence, we have very strong evidence for the fair coin hypothesis as against its alternative. If the fair coin hypothesis is in fact false, but we have strong evidence for it, then the probability of misleading positive evidence is only 0.000298 which is appreciably smaller than 0.1562 when we have weak evidence for the biased coin hypothesis. This shows that as evidence becomes strong the probability of misleading evidence goes down subject to some constraints, as in the tuberculosis case, from 0.1562 to 0.000298.

Examples like these show how, within our framework, the probability of misleading evidence can be made precise. In that framework, even though there is strong evidence for the presence of tuberculosis against its alternative and the inference is correct, it is possible that the disease is absent, in which case the positive test result is misleading.

There are two key theorems regarding the probability of misleading evidence that deserve mention. The first theorem states that in scenarios like this (comparison of simple hypotheses involving the tuberculosis example, the coin tossing example and the like), the probability of observing misleading evidence can never exceed $1/k$, where k indicates the strength of evidence represented by LR (Royall 1997, 2004). In the tuberculosis case, it is $1/26 \approx 0.038$ which is slightly higher than 0.0285. When sample size is sufficiently large, as the evidence becomes stronger, the probability of misleading evidence becomes lower subject to certain constraints[11] (Royall 1997, 2004). This bound, 1/k, on the probability of observing misleading evidence is a "universal bound" derived deductively from the probabilities of the data under these two competing models.

These results concur with Royall's general conclusion that when the sample size is very large, as evidence becomes strong the probability of misleading evidence decreases.[12] To be more precise, a second theorem concerning the probability of misleading evidence is that when the sample is very large, the probability of strong evidence for the correct hypothesis approaches one, whereas the probability for observing misleading evidence approaches zero (subject to certain constraints).

Summary

Our aim in this chapter was not to show that Achinstein's account of evidence is mistaken, no more than it was our aim in the preceding chapters to show that the subjective Bayesian, error-statistical, and selective confirmation accounts of evidence were wrong. More to our point is the fact that we differ in important ways and that, in particular, an account of evidence which is quantitative and comparative, provides in an intuitive way for misleading evidence, and for the reasons mentioned rejects the "true-model" assumption common to all of the other accounts discussed, provides, in our view, a more accurate and insightful description of current scientific practice.

It now remains to set out the ways in which the several accounts of evidence deal with the celebrated "paradoxes of confirmation," and then to apply our own account to a classical epistemological puzzle.

---

[11]There is a distinction not discussed here between a pre-data (1/k) and post-data (1/LR) probability of misleading evidence, (see Taper and Lele 2011). While acknowledging that the pre-data probability of misleading evidence was useful in conceptualization and study design, Royall rejected the use of the probability of misleading post-data. Post data, the evidence is either misleading or it is not; there is no probability involved. We modestly disagree with Royall on the utility of post data misleading evidence. First, there is no established scale for presenting evidence. The medical field often uses $\log_{10}(LR)$, ecology often uses $\ln(LR)$, and there are other formats. A post-data probability of misleading evidence, recognized as a counterfactual probability, not a belief-probability, can be useful in communicating the strength of evidence to scientists trained in classical error statistics. Second, as we have pointed out in Chap. 6, recognizing the implicit post-data error probability makes the relationship between evidential statistics and severe testing easier to understand. Finally, in some situations, such as evidence in the presence of nuisance parameters, or multiple comparisons, the post-data probability of misleading evidence may to some extent become uncoupled from the likelihood realization.

[12]Royall (2004).

References

Achinstein, P. (2010). Induction and severe testing: Mill's Sins or Mayo's error? In D. G. Mayo & A. Spanos (Eds.), *Error and inference* (pp. 170–188). Cambridge: Cambridge University Press.

Achinstein, P. (2001). *The book of evidence*. Oxford: Oxford University Press.

Achinstein, P. (1983). *The concept of evidence*. (ed.) Oxford: Oxford University Press.

Achinstein, P. (2005). *Scientific evidence*. Baltimore: The Johns Hopkins University Press.

Bandyopadhyay, P., Nelson, D., Greenwood, M., Brittan, G., & Berwald, J. (2011). The logic of Simpson's paradox. *Synthèse, 181*, 185–208.

Binford, S., & Binford, L. (Eds.). (1968). *New perspectives on archeology*. Chicago: Aldine Publishing Company.

Cartwright, N. (1983). *How the laws of physics lie*. Oxford: Clarendon Press.

Conee, E. (2004). Heeding misleading evidence. In E. Conee & R. Feldman (Eds.), *Evidentialism*. Oxford: Oxford University Press.

Longino, H. (1990). *Science as social knowledge*. Princeton: Princeton University Press.

Nye, M. (1980). N-rays: An episode in the history of science and psychology of science. *Historical Studies in the Physical Sciences, 11*, 125–156.

Royall, R. (1997). *Statistical evidence: A likelihood paradigm*. New York: Chapman Hall.

Royall, R. (2004). The likelihood paradigm for statistical evidence. In M. L. Taper & S. R. Lele (Eds.), *The nature of scientific evidence*. Chicago: University of Chicago Press.

Taper, M., & Lele, S. (2011). Evidence, evidence functions, and error-probabilities. In P. Bandyopadhyay & M. Forster (Eds.), *Handbook of statistics*. Amsterdam: Elsevier, North-Holland.

Part III
Applications of the Confirmation/Evidence Distinction to Epistemological Puzzles

Chapter 9
The Paradoxes of Confirmation

Abstract It is easy to resolve a contradiction. All you have to do is reject or reconfigure one of the premises of the argument that leads to it. What makes *paradoxes* so difficult to resolve is that the assumptions that generate them are so intuitive that they resist rejection or reconfiguration. The "paradoxes of confirmation" have been especially difficult to resolve. As much is indicated by the vast literature to which they have given rise. The "raven" and "grue" paradoxes are associated with, and often thought to cause problems for, the so-called "positive instance" account of confirmation. The "old evidence" paradox arises in connection with traditional Bayesian accounts of confirmation and, in the minds of some, is a decisive objection to it. These two accounts differ in a number of important ways. What they share is the assumption that the notions of confirmation and evidence are inter-definable, an assumption so deeply embedded that it has altogether escaped notice. Our object in this chapter is to show, once again, why confirmation and evidence should be distinguished, this time because their conflation is one root of the paradoxes. The work done by many others on the paradoxes, much of it technical, has thrown a great deal of light on our inductive practices. In providing a unified, if admittedly rather general treatment of them, we hope to indicate a new direction for this work.

Keywords Positive instance account of confirmation · Raven paradox · Grue paradox · Old evidence paradox

P.S. Bandyopadhyay et al., *Belief, Evidence, and Uncertainty*,
Philosophy of Science, DOI 10.1007/978-3-319-27772-1_9

The Positive Instance Account of Confirmation

On the positive instance account of confirmation,[1] the *direct* evidence for a hypothesis consists of its positive and negative instances. If an object *o* is *A* and also *B*, that is, satisfies the description "is *A* and *B*," then *o* is a *positive instance* of the generalized conditional "All A are B" (which is taken as representative of scientific hypotheses[2]). If an object *o* is *A* but not also *B*, then *o* is a negative instance of the hypothesis, *H*. To put this in the intended vocabulary, *o* confirms *H* just in case it is a positive instance of *H*.[3] In contrast, on the traditional Bayesian account, to confirm a hypothesis is to increase its posterior over its prior probability with respect to a set of data, to disconfirm a hypothesis is to lower it.

There are a variety of differences between these accounts. But what is crucial here is their similarity. For both assume that the evidence in favor of a hypothesis confirms it. Indeed, on both accounts, "confirmation" and "evidence" are inter-definable. On the one, a positive instance is simultaneously evidence and confirmatory, on the other, evidence is any data that raise the posterior probability of the hypothesis above its prior and hence confirm it. Confirmation of hypotheses takes place whenever we have rounded up enough evidence—positive instances or probability-raising data (whose description is often a consequence of the hypotheses)—for them.

The intuition underlying the positive instance account of confirmation is this: we test a generalization by looking at *samples* of it since, in the very nature of the case, it is impossible to *deduce* from the observations we might make that the generalization is true. Other things being equal, positive instances or samples confirm it; the more such samples we have observed, the more highly confirmed it is. This seems to be as much common sense as science. The point of paradox is to call common sense into question. The "raven paradox" suggests that we always have too much evidence on the positive instance account, the "grue paradox" that we have too much confirmation. Again, both paradoxes rest on a conflation of the two notions.

---

[1] The classic account is in Carl Hempel, "Studies in the Logic of Confirmation," in Hempel (1965).

[2] And thereby fails to include hypotheses that do not take this form, including the individual diagnostic hypotheses on which we have focused attention and, more generally and importantly, all statistical hypotheses.

[3] In Hempel's words, these are "the conditions under which a body of *evidence* can be said to *confirm* or *disconfirm* a hypothesis of empirical character." (Hempel 1965), p. 9 (our italics).

The Raven Paradox[4]

Suppose one seeks to confirm the hypothesis, H, that all ravens are black. On the positive instance account, one at least initially looks around for ravens; they constitute appropriate instances of the hypothesis and are positive if also black. But H is also logically equivalent to H', that all non-black things are non-ravens. So anything which is neither black nor a raven is a positive instance of, hence confirms H'. If it is required, as seems entirely reasonable, that objects confirming a hypothesis confirm all hypotheses logically equivalent to it—the "equivalence condition"—there emerges the paradoxical result that whatever is not black and not a raven, from white shoes to red herrings, also confirms the hypothesis that all ravens are black.

The feeling of discomfiture that these examples and others like them provoke is well expressed by Nelson Goodman. "The prospect of being able to investigate ornithological theories without going out in the rain is so attractive that we know there must be a catch in it."[5] Yet the implausible result that just about anything confirms the hypothesis that all ravens are black arises from some very natural assumptions.

Many different attempts have been made to resolve the raven paradox. It is not necessary to catalogue them here. In our view, a white shoe does not constitute *evidence* for the hypothesis that all ravens are black as against such logically equivalent generalizations that all non-black things are non-ravens since it fails to distinguish between the original generalization and any of its logical equivalents. But by the very same token, black ravens do not serve as evidence for the hypothesis that all ravens are black as against the hypothesis that all non-black objects are non-ravens. Since the two hypotheses are logically equivalent, they are equally likely on all and only the same data. In which case, no data can constitute "evidence" for the one and against the other.

Since this resolution of the paradox might seem both counter-intuitive and question-begging, not to say rushed, a word or two more is in order. Two conditions were taken to generate the Raven Paradox: that a generalization is confirmed by its positive instances and that a positive instance that confirms a generalization at the same time confirms any logically equivalent generalization. We think, in part for reasons already given, that the positive instance account of confirmation is not nearly so satisfactory as our own, typically Bayesian, positive relevance account, but our point now is not to dismiss or amend it as an account of confirmation. Nor do we want to reject the equivalence condition as Hempel states it; data that confirm a hypothesis should, on reasonable rationality requirements, confirm all of its logical equivalents. Logically equivalent propositions must always have the same probability. Moreover, these two conditions do not entail a formal contradiction, in which case we would have to reject at least one of them. The problem is, rather, to

---

[4]See Hempel (1965, pp. 14ff.)

[5]Goodman (1983, p. 70).

locate the source of the aura of paradox. We think that it lies in the conflation of confirmation with evidence,[6] that is, to identify data that confirm a hypothesis with evidence for it. The fact that an object is a positive instance of a hypothesis does not by itself entail that it is evidence for it.[7] Evidence as we have insisted from the outset should be understood as a comparative and contextual notion, in terms of the ratios of the data on competing hypotheses and not in terms of the positive instances of individual hypotheses. On this understanding, it is therefore misleading to assert that if D constitute evidence for a hypothesis H, then they constitute evidence for any logically equivalent hypothesis H'. For D constitute evidence for H only in a comparative context, when the likelihood of D on H is greater than the likelihood of D on another hypothesis. This is possible only if the two hypotheses being compared are not logically equivalent. Black ravens (or observations thereof) constitute evidence for the hypothesis that all ravens are black as against the hypothesis that some ravens are not black and many others, e.g., that ravens are red. If there is a problem with taking white shoes as evidence for the hypothesis that some ravens are not black, it is not so much that they can be ruled out on precise and principled grounds as that it is very difficult to know how to estimate the likelihood of their observation on the hypothesis that some ravens are not black.[8]

---

[6]Hempel's original discussion of the Raven Paradox in Sect. 5 of Hempel (1965) both makes explicit and depends on running confirmation and evidence together. "This implies that any non-raven represents confirming evidence for the hypothesis that all ravens are black" (p. 15, our italics).

[7]Like us, error-statisticians reject the idea that non-black non-ravens are evidence for the raven hypothesis, but for a very different reason: examining non-black non-ravens would not constitute a severe test of the raven hypothesis. The probability of finding non-black non-ravens on the raven hypothesis is very low. The test would have the same result if ravens came in a variety of colors. Error-statisticians would presumably also discount finding black ravens for the hypothesis that all ravens are black unless the testing procedures would be likely to turn up non-black non-ravens if the raven hypothesis were false. Among other things, this might take the form of gathering data adjusted for gender and geography. On our approach, black ravens do constitute evidence for the hypothesis that all ravens are black as against the hypothesis that they come in different colors. How strong this evidence might be is measured in terms of their respective likelihoods (which in a real-world test would include background information concerning gender and geography. See Giere (1970, p. 354).

[8]Royall in his commentary on the Raven Paradox (in an Appendix to his 1997) observes that how one got the white shoes is inferentially important. If you grabbed a non-raven object at random, then it does not bear on the question of whether all ravens are black. If on the other hand you grabbed a random non-black object, and it turned out to be a pair of shoes, then it provides a very tiny amount of evidence for the hypothesis that all ravens are black because the first sample estimates the proportion of non-black objects that are ravens as 0. Quantifying that evidence would be difficult without knowing how many non-black objects there are. In other words, the problem is one of confusing sample spaces. If you divide the world into two bins, ravens and non-ravens, no amount of sampling in the non-raven bin will give you any information about colors in the raven bin. However, if you divide the world into a bin of black things and a bin of non-black things, then finding a shoe in the non-black bin is evidence that all ravens are black (although fantastically weak evidence). The white shoe has increased the likelihood (albeit infinitesimally) that the proportion of items in the non-bland bin that are ravens is 0.

It is instructive, we believe, to look at Hempel's own way of understanding the aura of paradox generated by his account of confirmation. According to him, the trouble lies neither in taking samples nor in the equivalence condition, but in thinking that the result to which they lead is paradoxical. In his well-known words, "The impression of a paradoxical situation is not objectively founded; it is a psychological illusion."[9] He tries to wave it away.

In Hempel's opinion, this "illusion" has two sources. One is the mistaken tendency to think that hypotheses of the form "All R's are B's" are about R's only, that R's constitute their subject matter. When this hypothesis is paraphrased into a language more perspicuous than English, in particular into the language of first-order quantification theory, as $(x)(Rx \rightarrow Bx)$, it becomes clear that such hypotheses are really about all objects whatsoever within the range of the universal quantifier (x). But in his mistaken assumption that evidence can be identified with "positive instances," in terms of its syntactic character alone and as such, Hempel misses the point: it is not whether hypotheses are, on logical analysis, (really) "about" all objects whatsoever, but whether there is a statistically-relevant way in which to estimate the likelihood of data on them.[10] Without any further information, their observation is equally likely on the hypotheses that all ravens are black and that no ravens are black.

The other source of the "illusion" that his account of confirmation has paradoxical consequences is, Hempel says, the mistaken tendency to take so-called "background information" into account. What leads one to rule out red herrings as acceptable instances of the raven hypothesis is the fact that one *already* knows that they are neither ravens nor black. Hence they do not provide *added* support for it. But if any given object were referred to simply as "object x," and specified no further, then the discovery that it was neither a raven nor black would confirm the hypothesis that all ravens are black, although perhaps not to the same degree as a black raven would, even if the test object happens to be a red herring.

Hempel's first point has to do with "evidence," the second with "confirmation." The first does not at all accord with scientific practice, nor is any reason given why practice should in this case bow to a very abstract theory. More than consistency is required for the data to "fit" or be relevant to a hypothesis. As for the second, there is no reason to deny that our confidence in particular hypotheses will vary as a function of background information; on the Bayesian account, for example,[11] since

---

[9](Hempel 1965, p. 18).

[10]It needs to be emphasized that our characterization of evidence, quite apart from the requisite distinction from confirmation that it makes possible, rules out ab initio any attempt to characterize "evidence" as such, for example in terms of the form of the sentences which represent it (as in Hempel) or its alleged incorrigibility or its "givenness" (as in Lewis), and on. Evidence has no intrinsic characteristic which identifies it, but is constituted by context-relative data whose strength is measured by the likelihood ratio (or something comparable).

[11]See Good (1960, pp. 149ff.) for the classic "Bayesian analysis," but also Chihara (1987), who makes the case for a variety of "Bayesian" analyses. For a more recent and very insightful discussion, see Fitelson and Hawthorne (2006).

the size of the black raven reference class is much smaller than the size of the non-black non-raven reference class, the observation of a black raven should raise the probability of the "all ravens are black" generalization to a much greater degree than will the observation of a white shoe, and, for similar reasons, the likelihood of spotting a raven which is also black on this generalization is much greater than the likelihood of finding a red herring on the denial of this hypothesis, viz. "some ravens are not black."[12] No one excludes background information in practice, and again no reason is given why she should do so. More to our point is that the claim that white shoes in the absence of background information add to the confirmation of the raven hypothesis, even if true, does not entail that the paradox is a "psychological illusion." The illusion has a conceptual source, the confusion of confirmation with evidence.[13]

The Grue Paradox

The "grue paradox." is at least as well known as the "raven." But it is worth setting up in the usual way before proceeding.

Suppose that all emeralds examined before a certain time t (say, the year 2020) are green. On the positive instance account of confirmation, positive instances of the hypothesis that all emeralds are green, in this case green emeralds, confirm it. Or as the inventor of the paradox, Nelson Goodman, puts it more precisely: "Our *evidence statements* assert that emerald a is green, that emerald b is green, and so on; and *each confirms the general hypothesis* that all emeralds are green. So far, so good."[14]

Goodman then introduces a new predicate, "grue," which "applies to all things examined before t just in case they are green or to others just in case they are blue." Now consider the two hypotheses:

1. H_1: All emeralds are green.
2. H_2: All emeralds are grue.

It should be clear from the way in which the predicate "grue" was introduced that at time t all the apparent evidence for H_1 is also evidence for H_2, and vice versa. They are equally well confirmed; for at time t, the two hypotheses have exactly the same positive instances. But this is paradoxical.

---

[12]We earlier indicated the necessity of including background information in our calculation of posterior probabilities and likelihoods, but since it had little bearing on the main theme of the monograph, have not made much of it. When one gets to the details of the applications of most of the main theories of confirmation and evidence to particular cases, background information is very important, although difficult to delimit in a principled way.

[13]Again see Royal (1997).

[14](Goodman 1983), p. 74 (our italics).

In the first place, although equally well confirmed by their positive instances, they license incompatible predictions about emeralds examined after t. This by itself is not paradoxical; at least since Leibniz, it has been generally known that through a finite number of (data-) points, an infinite number of curves could be drawn. But the fact that all the emeralds examined so far have been green—hence also grue—seems not in the least to support the prediction by way of H_2 that the next emerald examined after t will be blue, although it does seem to support the prediction via H_1 that it will be blue. One way to put the intuition here is to say that we have no plausibility argument available that would provide grounds on the basis of which we could expect that it will be blue after t, no statistical distribution that would countenance its possibility or provide for a perceptual error of this magnitude.

In the second place, "grue" is an arbitrary predicate; there is no more reason for thinking that emeralds examined after time t will be blue than for thinking that they will be red. So there is no more reason for asserting "All emeralds are grue" than there is for asserting "All emeralds are gred." We can cook up any number of "grue-type" predicates. All will be true of emeralds to the same extent that "green" is, for the hypotheses in which they figure would seem to be supported by precisely the same evidence. The evidence supports just about any hypothesis we wish to frame about the colors of emeralds over time. And this result is, in Goodman's word, "intolerable."

It is tempting to object immediately that "grue" and "green" are unlike in a crucial respect. The meaning of "grue" but not "green" includes reference to a particular time t, and hence the hypotheses in which it figures lack the generality characteristic of physical laws. This objection can be dismissed, however. Suppose we have the predicate "bleen;" an object is bleen just in case it is blue before time t and otherwise green. In which case, the meaning of "green" just as much as that of "grue" includes a reference to a particular time, grue before t, otherwise bleen. If the apparently uncaused change from green to blue at t seems adventitious, so too must a similar change from grue to bleen.

Any number of attempts have been made to solve the grue paradox. Most of them focus on apparent asymmetries between the hypotheses in which "green" and "grue" figure with respect to the evidence.[15] For the moment it is enough to assert without providing arguments that whether they invoke "natural kinds," or the law of large numbers, or an innate ordering of hypotheses, or ostensive definability, all fail. There simply are no fundamental differences—syntactic, semantic, even

---

[15]In his "Forward" to Goodman (1983, p. ix), Hilary Putnam simply lays it down that "in order to 'solve' Goodman's problem one has therefore to provide some principle capable of selecting among inferences that do not differ in logical form, that is, on the basis of certain predicates those inferences contain." It is revealing that Putnam himself rejects the leading solutions along these lines, is wary of Goodman's own, and has nothing better to suggest. Our solution, in sharp contrast, is not to look at the predicates, but at the structure of the inferences.

pragmatic—on the basis of which we can distinguish "grue" from "green" with respect to their confirmability.[16]

Another popular tack is to reconfigure or abandon the "positive instance" account of confirmation that figures as an important premise in Goodman's argument. Thus it is maintained that if we put constraints on what is to count as a "positive instance," or take consequences rather than positive instances of hypotheses as confirming them, or develop a quantitative account of confirmation, then the paradox will be blocked. But the "positive instance" account is itself not to blame; we know of no otherwise plausible confirmation theory in which, on the present assumptions, some version of the grue paradox cannot be formulated.

These failures, and others like them, provide inductive evidence that the paradox has no effective resolution, a claim that several philosophers have been ready to accept.[17] In their view, in fact, the attempt to rule "grue" inadmissible on a priori

---

[16]Achinstein's attempted resolution of the paradox (Achinstein 2001) is both typical and instructive. His argument proceeds in three steps:

1. If D is to provide evidence for H, then the posterior probability of H given D must be high;
2. If the probability of D given H is high, then D must be appropriately varied;
3. In the case of the grue hypothesis, the green emerald data are not appropriately varied.

Therefore, …

…There are problems with the second premise, that high posterior probability requires appropriately varied evidence. This requirement does not follow from the rules of probability theory, nor is it easy to see how variety of evidence (as we emphasized in the preceding chapter), however laudable a methodological goal, is to be included in a formal account except in terms of something like a pair-wise comparison of hypotheses. But the third premise is for present purposes more problematic. Achinstein argues for it as follows. "Grue" is a disjunctive predicate; it applies to emeralds examined before some arbitrary date if they are green and after said date if they are blue. "Appropriately varied" data would therefore include observation of emeralds both before and after. But it is a premise of the argument that "before" data only are taken into consideration. Hence they cannot constitute "evidence" in Achinstein's sense for the hypothesis that all emeralds are grue. The same line of argument would apply to all similarly disjunctive predicates, but not to such atomic predicates as "green," the explanation of why the latter but not the former are entrenched in our verbal and experimental practice. Goodman would undoubtedly reply that "green" can also be construed as a disjunctive predicate, grue before t, bleen after it. If we were to protest that this way of speaking does not register the fact that a change from green to blue would have to be *caused*, whereas a "change" from grue to bleen would require no cause, Goodman could point to Hume's analysis, on which "causes" are no more than habits and up-date it to *linguistic* habits. Moreover, we follow Rosenkrantz (see Footnote 11, Chap.8) in thinking that there are no good reasons for excluding disjunctive or "bent" predicates other than to avoid the Grue Paradox, an ad hoc and unnecessarily conservative move.

[17]At least at one point in his career, and as expressed in correspondence with one of the authors, Glymour drew the conclusion, as did Goodman, that confirmation is relative to a particular interpretation of the syntax in which hypotheses and data are formalized, or, equivalently, to a particular language. On his view at one time, this conclusion can be extended to any formal confirmation theory, whether it be positive instance, Bayesian, or "bootstrap." The moral, apparently, was that the Grue Paradox cannot be used to decide between existing confirmation theories. None of them can solve it without adding certain presuppositions about meaning, a view which echoes that of Carnap.

grounds is inimical to the "no holds barred" sort of inquiry characteristic of science, the most important presupposition of its progress.[18]

We agree that there are no grounds on which "grue" can be ruled out in advance. But this laudable pragmatism is consistent with maintaining that there is, nonetheless, something problematic about the paradox in which "grue" figures. What is problematic, in our view, is the idea that, whatever the hypothesis, there is an arbitrary and incompatible alternative to it that is confirmed by precisely the same evidence.

It is precisely this last phrase, "confirmed by the same evidence," that is the source of the problem. It isn't the weirdness of "grue." It isn't the use of a particular confirmation theory. Rather, the source of the problem has to do with the way in which virtually all confirmation theories are stated, that a hypothesis' positive instances or consequences or what have you are evidence for and *hence* confirm it, or the other way around, that what confirms a hypothesis is *hence* evidence for it. The source of the problem is a well-nigh irresistible tendency to run "confirmation" and "evidence" together.

On certain plausible assumptions, the observation of green emeralds raises the probability of both the "grue" and "green" hypotheses about emeralds on the Bayesian account, and in this sense confirms them.[19] But since green emeralds do not serve for now to distinguish the two hypotheses, then green emeralds do not provide "evidence" for either of them in what we have argued is a basic and intuitive sense of the term. In our statements of the paradox to this point, we have said that the two hypotheses were supported by exactly the same (actual)

---

[18]See Rosenkrantz (1981), Chap. 7, Sect. 1. It has been proposed several times, for example, that Newton's law of universal gravitation be amended; the force between two objects varies inversely as the *square* of the distance between them, up to some distance d. But for distances greater than d, the force varies inversely as the *cube* of the distance between them. Such proposals to "correct" Newton's law were made by the Royal Astronomer G.B. Airy (1801–1892) in the attempt to explain observed irregularities in the motion of the planet Uranus, and again in the 20th century by the physicist Hugo von Seeliger (1849–1924) in the attempt to make the mean density of the university everywhere constant (the inverse square law implies a concentration of matter around centers of maximum density).

[19]As soon as the background information that emeralds don't undergo sudden (and otherwise uncaused) color changes is factored in, then the posterior probability of the green emerald hypothesis is much higher. This background information is easily incorporated into the relevant priors; we accord the green emerald hypothesis a higher degree of belief on our knowledge of how the world works (i.e., acceptance of the grue emerald hypothesis would very much complicate our current scientific picture). According to Goodman, on the green emerald hypothesis, too, emeralds undergo sudden (and otherwise unexplained) color changes, from "grue" to "bleen." So we can't allow our views about color "changes" to figure in our confirmatory accounts. But (a) the green/grue change implicates changes in many other parts of physical theory than does the grue/bleen "change," and (b) Goodman's own view suggests that our relative familiarity with "green," its greater degree of "entrenchment" in our verbal and other practices, will lead people to accord the green emerald hypothesis a higher degree of initial belief. In either case, the green emerald case is more highly confirmed. But the two hypotheses have, at least before t, the same evidence class. If one assumes that a hypothesis is confirmed to the extent that there is evidence for it, then paradox ensues. To avoid the paradox, we simply drop the assumption.

evidence.[20] But although discovery of green emeralds does, we might grant, increase the probability of one or the other of the two hypotheses, it does not constitute evidence for one or the other. But if it does not constitute evidence, then the paradoxical idea that exactly the same evidence could support incompatible theses is undermined from the outset.[21]

We could put the point the other way around. The grue hypothesis is "arbitrary" in just this sense, that it is so framed that no available evidence *could* distinguish it from its green partner. But where no distinction on the basis of the data is possible, we have no evidence, for *evidence* tells for and against particular hypotheses.[22] Likelihood ratios are introduced to make this idea explicit: if the ratio of the likelihoods of the data on two hypotheses is no greater than 1, then the data adduced for either do not constitute evidence for it.

It is often held that Goodman's paradox shows conclusively that (as against Hempel and others) one cannot give a purely syntactical account of confirmation. This claim does not cohere with Goodman's own statement of its "residual problem:" which (among the innumerable "grue" type variations on them) hypotheses are (really) confirmed by their positive instances, since "positive instance" is itself a syntactic notion. It also misleads. For it suggests that we must press on to find a semantic or, eventually, "pragmatic" account of confirmation. The fact of the matter, as we have tried to make both clear and persuasive, is that the notion of evidence cannot be captured by positive instances or consequences of hypotheses, or any other general and non-contextual attribute, regardless of how it is characterized. Evidence does not have some intrinsic characteristic which identifies it as

---

[20]See Goodman (1983), p. 74: "Then at time *t* we have, for each *evidence statement* asserting that a given emerald is green a parallel *evidence statement* asserting that the emerald is grue" (our italics).

[21]It might be objected that our way with the paradox is *ad hominem*. That is, although *Goodman* conflates confirmation and evidence, we could restate the paradox in terms of confirmation alone: incompatible hypotheses are confirmed by the same *data*, observations of green emeralds, whether or not these data are said to constitute "evidence" for either one. This objection presupposes, what both we and Goodman deny, that the two hypotheses have the same priors; indeed, it goes through only if we assume that the "positive instance" account is correct. At the same time, it is hardly surprising, much less paradoxical, that incompatible hypotheses might be confirmed by the same data. This fact has long been known and gives rise to the very difficult "curve-fitting problem" at the foundations of statistical inference. The real bite comes when we begin, as with Goodman, to speak of "evidence," for this suggests a way of distinguishing between other indistinguishable hypotheses.

[22]Thus green emeralds constitute evidence for the hypothesis that all emeralds are green vis-à-vis the hypothesis that all emeralds are bleen, but do not constitute evidence for the green hypothesis vis-à-vis the hypothesis that all emeralds are grue. The likelihood of green emeralds (examined before some arbitrary time *t*) on the green hypothesis is vastly greater than the likelihood of green emeralds on the bleen hypothesis, whereas it is the same on both green and grue hypotheses.

such. Rather, it is to be understood in functional terms, as that which allows us to distinguish between hypotheses on the basis of our data.[23]

The Old Evidence Paradox

The "old evidence" paradox raises problems not for positive instance, but for Bayesian accounts of confirmation.[24]

The classic formulation of the old evidence problem is due to Glymour.[25] Many theories come to be accepted not only because they yield novel predictions that are subsequently verified, but because they also account more successfully than their competitors for observations long since made. Copernicus, for example, supported his heliocentric theory in part with observations dating back to Ptolemy. A principal support of Newton's theory of universal gravitation was his derivation of the laws of planetary motion already established by Kepler. But on the Bayesian account, this sort of "old" evidence apparently does not confirm new hypotheses, a fact that makes hash of the history of science. For suppose that data D are already known when hypothesis H is introduced at time t. If D are known, according to the old evidence problem formulation, then $\Pr(D) = 1$. If D is a logical consequence of H, then the likelihood of D given H, $\Pr(D \mid H)$, is also 1. Thus by Bayes Theorem, $\Pr(H \mid D) = \Pr(H) \times 1/1 = \Pr(H)$. That is, the posterior probability of H given D is the *same* as the prior probability of H; D does not raise its posterior probability, hence, contrary to practice and intuition, does not confirm it.[26]

It should be clear how the old evidence paradox rests on a conflation of "evidence" with "confirmation" typical of philosophical work on the topic of confirmation generally, and more particularly work done by Hempelians and subjective Bayesians. Thus we are told by Bayesians that D is evidence for H if and only if $\Pr(H \mid D) > \Pr(H)$, where the latter probability is just an ideal agent's current belief probability distribution. Once this conflation is undone, by distinguishing sharply between evidence and confirmation, then so too is the paradox. For Glymour argues from the fact that in cases of "old" data "the conditional probability of T [i.e., the hypothesis] on e [the datum] is therefore the same as the prior probability of T" to

---

[23]That the intuition involved in our account, data count as evidence just in case they serve to distinguish hypotheses, has a long history is indicated by William Caxton, in his *Deser Eng.* of 1480: "He maketh no evidence for in neyther side he telleth what moveth him for to saye."

[24]Although some prominent Bayesians, particularly those of an "objectivist" orientation, maintain that the air of paradox is illusory. See, for example, Roger Rosenkrantz, "Why Glymour is a Bayesian," in Earman (1983), especially pp. 85-6. In the same volume, Daniel Garber, "Old Evidence and New," essays a "subjectivist" attempt to disarm the problem. See Bandyopadhyay (2002) for reasons why the Bayesian account of confirmation cannot, on either of its standard variants, solve the old evidence problem.

[25]See Glymour (1980), Chap. III.

[26]The argument can be re-cast in such a way that it does not depend on $\Pr(D) = 1$.

the conclusion that "*e* cannot constitute evidence for *T*,"[27] and this conclusion can now be seen to be a *non-sequitur*. "Old" or "new" evidence, evidence is evidence, an intuition that our non-Bayesian conception of it captures. On our account, data provide evidence for one hypothesis against its alternative if the likelihood ratio is greater than one. If the latter condition holds then data do provide evidence for one hypothesis against its alternative irrespective of whether they are old or new. We will revisit this point at the end of our discussion of the old evidence problem.

It follows as a corollary of this account that evidence is agent-independent. On Glymour's formulation of the old evidence problem, whenever *D* confirm *H*, they become evidence for *H* relative to an agent *A*, in which case whether *D* is evidence for *H* depends on whether *A* knows or believes *D* to be the case. But as we have insisted from the outset, whether *D* is evidence for *H* has nothing to do with whatever *A* believes about either *D* or *H*. As our paradigm tuberculosis diagnostic example also illustrates, *D* can provide evidence for *H* independent of what an agent knows or believes about *D* or *H*.

Perhaps the most celebrated case in the history of science in which old data have been used to justify a new theory concerns the perihelion shift of the planet Mercury (*M*) and the General Theory of Relativity. Of the three classical tests of GTR,[28] *M* is regarded as providing the best evidence.[29] According to Glymour, however, a Bayesian account fails to explain why *M* should be regarded as evidence for GTR. For Einstein, $Pr(M) = 1$, since *M* was known to be an anomaly for Newton's theory long before GTR came into being.[30] Einstein derived *M* from GTR; therefore, $Pr(M \mid GTR) = 1$. Once again, since the conditional probability of GTR given *M* is the same as the prior probability of GTR, it follows on the Bayesian account that *M* cannot constitute evidence for GTR. But given the crucial importance of *M* in the acceptance of GTR, this is at the very least paradoxical. It is known, moreover, that the old evidence problem continues to haunt the Bayesian account of evidence even if the probability of the data is not equal to, but is close to one. This fact, however, does not bear on our resolution, that is, whether or not the problem is premised on the probability of data equal to or close to one, it still conflates in a Bayesian way the concepts of "evidence" and "confirmation" that we have gone to great lengths to distinguish. This is the core of our solution.

---

[27]*Ibid.*, p. 86.

[28]Summarized in Chap. 6.

[29]See Brush, "Prediction and Theory Evaluation: the Case of Light Bending," *Science*, 246 (1989), pp. 1124-129; Earman and Janssen, "Einstein's Explanation of the Motion of Mercury's Perihelion," in Earman, et al., eds., *The Attraction of Gravitation* (Cambridge, MA: MIT Press 1993); Roseveare, *Mercury's Perihelion from LeVerrier to Einstein* (Oxford: Oxford University Press 1982). We are especially indebted to Earman's account of the tests of the GTR.

[30]For our purposes it is not necessary to decide any of the historically delicate questions concerning what Einstein knew and when he knew it; what he knew or didn't know at the time of his discovery of GTR has nothing to do, as against Glymour's paradox, with the evidential significance of *M*. In trying to determine what constitutes "new" evidence, Imre Lakatos and Eli Zahar make a rather desperate appeal to what the scientist "consciously" knew when he came up with GTR. See their (1975).

Another way of approaching the problem is to claim that the old evidence problem should not really arise. According to this approach, its canonization in the philosophical literature results from a failure to note the difference between a random variable and its realization. A random variable can be defined "as a variable that takes on its values by chance."[31] A realization is an observation of one of those chance values. From a philosophical perspective, the confusion embodied in the old evidence problem stems from mistakenly identifying "knowing or observing the data" with "the probability of the data". These are indeed two different beasts. Consider the example of throwing a die. A roll may be mathematized as a random variable which can take on integer values from 1 to 6. Random variables are generally denoted by Roman capital letters, e.g. X, and their realizations by lower case Roman letters, e.g. x. If one throws a fair die and sees that a face with three pips is up, it is true that the probability is one that the throw scored a three. This is what we mean by "knowing or observing the data." However, the probability of seeing three pips on the next roll is still just one sixth. Observing a realization does not change the probability distribution of the random variable. The term $\Pr(D|H_i)$ in the numerator of Bayes' rule does not represent the scientist's belief that an observation of type d has occurred, but is instead the inherent propensity of the data generating mechanism embodied in the hypothesis, H_i, to produce data of type d (see Chap. 2, section on "Our Two Accounts and Interpretations of Probability" for more on interpretational issues). Technically, the term should be written as $\Pr(D = d|H_i)$, that is, the probability that the random variable D would take the observed value of d given the hypothesis H_i.[32] The notation $\Pr(d|H_i)$ is a shorthand used by statisticians on the assumption that no misunderstanding will be generated. Unfortunately, in this case, that assumption has been violated. Similarly $\Pr(d)$ in the denominator is also not the scientist's belief that she has observed data d, but is instead the marginal probability of data of type d being generated by the set of alternative models, $\Pr(D = d) = \sum_i (\Pr(D = d|H_i) \cdot \Pr(H_i))$. The prior probabilities, $\Pr(H_i)$, are subjective belief measures. Again, observing data d does not drag this probability, $\Pr(d)$, to one. So the old evidence paradox on this reading is simply an artifact of misconstruing knowing or observing the data as the probability of the data, but, as we have argued, they are two different concepts. So, we are able to dissolve the old evidence problem by making "a random/realized variable" distinction as well as by making one between "evidence" and "confirmation."

Some readers might wonder whether we have yet addressed the underlying and genuinely significant question, "do *old* data provide *evidence* for a *new* theory"? We will now address this more-focused question with the help of our account of evidence, leaving out further discussion of the distinction between random and realized variables.

On our evidentialist account M (i.e., the observed shift of Mercury) does constitute evidence, indeed, very significant evidence for the GTR. Consider GTR and

---

[31]See Taylor and Karlin (1998, p. 7).
[32]See Pawitan (2001, p. 427).

Newton's theory, NT, relative to M with different auxiliary assumptions for the two theories. Two reasonable background assumptions for GTR are (i) the mass of the Earth is small in comparison with that of the Sun, so that the Earth can be treated as a test body in the Sun's gravitational field, and (ii) the effects of the other planets on the Earth's orbit are negligible.[33] Let A_E represent those assumptions.

For Newton, the auxiliary assumption is that there are no masses other than the known planets that could account for the perihelion shift. Let A_N stand for Newton's assumption. We now apply our evidential criterion, the Likelihood Ratio, to a comparison of the two theories, albeit in a very schematic way. $\Pr(M \mid GTR \,\&\, A_E) \approx 1$, whereas $\Pr(M \mid NT \,\&\, A_N) \approx 0$. The LR between the two theories on the data approaches infinity, which is to say that M provides a very great deal of evidence indeed for GTR and virtually none for Newton's theory.[34] It is often held that, whatever the evidential situation, theories once accepted are not rejected except insofar as a better theory is available. But our way with "evidence" makes precise why this should be the case. Perturbations in the orbit of Mercury could not count as *evidence* against Newton's theory until there was a rival theory on which these perturbations were more likely. It is not that we do not want to leave a ship, however much it might be sinking, unless another is there to take us on board, but that in the absence of a comparison between competing hypotheses, we cannot know that we are sinking. Of course, there was a good deal of evidence for Newton's theory vis-à-vis its Aristotelian and Cartesian rivals.

The "old evidence paradox" can be dissolved; it rests on the mistaken assumption that evidence and confirmation are to be understood in exactly the same Bayesian terms. But there are two residual and very genuine problems concerning "old evidence" that might be mentioned since both have important implications for the development of an adequate theory of statistical inference.

The first problem has to do with *post hoc* data fitting. On our view, evidence is a deductive relationship between the data and the models under comparison; it is of no consequence when the data were collected or the models constructed. But on a finer-grained analysis, statistical evidence acts to inform scientific inference in some ways that are intuitively more credible than others. Notably, it is always possible to construct theories and models that provide a good fit to *any* given data set. It is just a matter of adding more and more parameters to the model.[35] No one would claim that data that had been fabricated to match a theory constitute evidence for the theory even though on the basis of the fabricated data the statistical evidence for that theory (over another) may be quite high. But to craft a theory to fit data after-the-fact is to engage in a roughly similar activity.[36]

---

[33]We owe this formulation of the background assumptions for both GTR and Newton's theory to John Earman in an email communication.

[34]See also Lange (1999) for a different approach to the "old evidence" problem..

[35]In the limit case, adding as many parameters as there are data points.

[36]Careful statisticians do post hoc data analysis all the time, but they label it as such and consider their results more as hypothesis-generating than as evidence-supporting.

Because of the complexities of ecological and evolutionary processes and the extreme difficulty of collecting data, ecologists and evolutionists have a tendency to explain whatever data are available in adaptionist terms, without any further constraints on the explanations than they conform generally to a Darwinian template. In a famous paper, "The Spandrels of San Marco and the Panglossian Paradigm: A Critique of the Adaptionist Program"[37] Stephen J, Gould and Richard Lewontin identified this as a tendency to tell Kipling-style "just so" stories, stories designed to explain the known facts but with little methodological or theoretical constraint on the explanations.[38] The stories are not worthless; for one thing, they may suggest new lines of inquiry or the search for additional data. But a commonly-held intuition is that data have more epistemic value with respect to hypotheses that were not designed to capture or cannot explain more than a given set of data than those that were and can.

Although Einstein's GTR was developed after Mercury's perihelion was known, it was not designed to predict the arc shift and, more importantly, it could explain and be tested with respect to a variety of new data. It is just that although the arc shift constituted "evidence," on our account as well as in the scientific community, new data constituted "better" evidence for it. We have not yet developed an account of the way in which this intuition is precisely to be captured.

The second, related, and so far unresolved "old evidence" problem is that the accommodation of a theory by the data might run the risk of over-fitting error (Hitchcock and Sober 2004). In model selection, over-fitting error occurs when a model describes random errors or noise instead of the underlying relationship between variables. This error arises when a model possibly contains to many parameters relative to number of observations. This over-fitting could arise in our evidential framework which can be handled by information criteria such as the Akaike Information Criterion (see the Appendix at the end of the monograph for how this over-fitting can be addressed using a real world example from ecology).

Our resolution of the old evidence problem turns on twin distinctions between evidence and confirmation, on the one hand, and randomness and realization, on the other. Conflating evidence and confirmation leads to the classical, or "Bayesian" form of the problem. Conflating randomness and realization leads to an ambiguity in the interpretation of likelihood that will infect any theory of statistical inference that makes use of them. Once likelihoods are correctly interpreted, and evidence understood in terms of a ratio between them, then the more-specific form of the evidence problem, "when do *old* data provide *evidence* for *new* theories?" can be answered without ambiguity and in a straightforward way: not when they further "confirm" a theory but when they discriminate between a new theory and an older rival and pronounce in favor of the former. But, our way with Glymour's paradox

---

[37]Gould and Lewontin (1979).

[38]This was not the end of the matter. Among the most interesting of the many papers critical of the Gould-Lewontin thesis is Dennett (1983).

does not yet provide a fully adequate account of the inferential force that old data provide new theories or of the over-fitting problem, issues that arise for every account of statistical inference.

Summary

We can summarize the situation as follows. In the raven and grue paradoxes, data intuitively fail to confirm some hypotheses, yet on the positive instance account must apparently be taken as evidence for them, while in the old evidence paradox, data intuitively succeed in supporting some hypotheses, yet on the Bayesian account must apparently not count as evidence for them. It undoubtedly seems naïve and unbecomingly immodest to resolve all three paradoxes with no more than a simple and straightforward distinction between evidence and confirmation or justification.Understood. But if our reading of the situation is correct, the paradoxes have a common thread, the confusion of evidence and confirmation/justification, and a single sharp stroke of the sword suffices to undo at least one of the knots into which they have become entangled over the years. Our hope is that all of the patient work devoted to picking them apart will now be seen in a new and we trust revealing light.

References

Achinstein, P. (2001). *The book of evidence*. Oxford: Oxford University Press.
Bandyopadhyay, P. (2002). The old evidence problem and beyond, given at the American Philosophical Eastern Division Meetings.
Brush, S. (1989). Prediction and theory evaluation: The case of light bending. *Science, 246*, 1124–1129.
Chihara, C. (1987). Some problems for Bayesian confirmation theory. *British Journal for the Philosophy of Science, 38*, 551–560.
Dennett, D. (1983). Intentional systems in cognitive ethology: The 'panglossian paradigm' defended. *The Behavioral and Brain Sciences, 6*, 343–390.
Earman, J. (1983). *Testing scientific theories* (Vol. 10, Minnesota Studies in the Philosophy of Science). Minneapolis: University of Minnesota Press.
Earman, J., & Janssen, M. (1993). Einstein's explanation of the motion of mercury's perihelion. In J. Earman, M. Janssen, & J. Norton (Eds.), *The attraction of gravitation*. Boston: Birkhäuser.
Fitelson, B., & Hawthorne, J. (2006). How Bayesian confirmation theory handles the paradox of the ravens. In E. Eells & J. Fetzer (Eds.), *The Place of Probability in Science* (Boston Studies in the Philosophy of Science 284). Dordrecht: Springer.
Garber, D. (1983). Old evidence and logical omniscience in Bayesian confirmation theory. In (Earman, 1983).
Giere, R. (1970). An orthodox statistical resolution of the paradox of confirmation. *Philosophy of Science 37*, 354−362
Glymour, C. (1980). *Theory and evidence*. Princeton: Princeton University Press.

Good, I. (1960). The paradox of confirmation. *British Journal for the Philosophy of Science, 11,* 145–149.

Goodman, N. (1983). *Fact, fiction, and forecast.* Cambridge, MA: Harvard University Press.

Gould, S., & Lewontin, R. (1979). The spandrels of San Marco and the Panglossian paradigm: A critique of the adaptationist program. *Proceedings of the royal society of London, 205,* 581–598.

Hempel, C. (1965). *Aspects of scientific explanation.* New York: Free Press.

Hitchcock, C., & Sober, N. (2004). Prediction vs. accommodation and the risk of over-fitting. *British Journal for the Philosophy of Science, 55,* 1–34.

Lakatos, I., & Zahar, E. (1975). Why did Copernicus' research programme supersede Ptolemy's? In R. Westfall (Ed.), *The Copernican achievement.* Berkeley: University of California Press.

Lange, M. (1999). Calibration and the epistemological role of conditionalization. *Journal of Philosophy, 96*(1), 294–324.

Pawitan, Y. (2001). *In all likelihood: modeling and inference using likelihood.* Oxford: Oxford University Press.

Putnam, H. (1983). Forward. In (Goodman, 1983).

Rosenkrantz, R. (1981). *Foundations and applications of inductive probability.* Atascadero, CA: Ridgeview Publishing.

Rosenkrantz, R. (1983). *Why Glymour is a Bayesian.* In (Earman, 1983).

Roseveare, N. (1982). *Mercury's perihelion from LeVerrier to Einstein.* Oxford: Oxford University Press.

Royall, R. (1997). *Statistical evidence: A likelihood paradigm.* New York: Chapman Hall.

Taylor, H., & Karlin, S. (1998). *An Introduction to stochastic modeling* (3rd ed.). San Diego: Academic Press.

Chapter 10
Descartes' Argument from Dreaming and the Problem of Underdetermination

Abstract Very possibly the most famously intractable epistemological conundrum in the history of modern western philosophy is Descartes' argument from dreaming. It seems to support in an irrefutable way a radical scepticism about the existence of a physical world existing independent of our sense-experience. But this argument as well as those we discussed in the last chapter and many others of the same kind rest on a conflation of evidence and confirmation: since the paradoxical or sceptical hypothesis has as much "evidence" going for it as the conventional or commonly accepted hypothesis, it is equally well supported by the data and there is nothing to choose between them. By this time, however, we understand very well that data that fail to discriminate hypotheses do not constitute "evidence" for any of them, i.e., that "data" and "evidence" are not interchangeable notions, that it does not follow from the fact that there is strong evidence for a hypothesis against one or more of its competitors that it is therefore highly confirmed, and that it does not follow from the fact that a hypothesis is highly confirmed that there is strong evidence for it against its rivals.

Keywords Argument from dreaming · Plausibility arguments · Thought-experiments · Under-determination

The Argument from Dreaming

Descartes sets the Dreaming Argument out very briefly in the first *Meditation*:

> How often, asleep at night, am I convinced of just such similar events—that I am here in my dressing gown, sitting by the fire—when in fact I am lying undressed in bed! Yet at the moment my eyes are certainly wide awake when I look at the piece of paper; I shake my head and it is not asleep; as I stretch out and feel my hand I do so deliberately, and I know what I am doing. All this would not happen with such distinctness to some asleep. Indeed! As if I did not remember other occasions when I have been tricked by exactly similar thoughts while asleep! As I think about this more carefully, I see plainly that there are never any sure signs by means of which being awake can be distinguished from being asleep.[1]

---

[1]Descartes (1984, p. 14).

© The Author(s) 2016
P.S. Bandyopadhyay et al., *Belief, Evidence, and Uncertainty*,
Philosophy of Science, DOI 10.1007/978-3-319-27772-1_10

But if, Descartes goes on to suggest, "there are never any sure signs by means of which being awake can be distinguished from being asleep", then there is no way in which we can be assured that any of the claims he makes about the world are true, since the only evidence on the basis of which he could make them is as much available asleep as awake, and asleep the claims are false.[2] Thus radical skepticism.

A great variety of attempts to undermine this argument have been made, none of them generally acknowledged as successful. Indeed, in the course of an extended treatment of it, Barry Stroud comes to the conclusion that the argument is sound; it cannot be undermined, only better understood.[3] We disagree.

There Is No Evidence that We Are Awake or Asleep

The first thing to note is that the dream argument rules out the possibility of "evidence" that we are awake. That is, the data in question, the various "sensible experiences" that we have, are equally likely whether we are awake or asleep. The ratio of their likelihoods is 1, in which case, on our characterization, the data do not provide evidence for one hypothesis or the other, that we are awake or asleep. It is in this sense, "proof" being supposed interchangeable with "evidence", that we cannot *prove* that we are awake.[4]

For many commentators on the dreaming argument, this is the end of the matter. As Stroud puts it, in light of the possibility that from one moment to the next we are dreaming, "our sensory experience gives us no basis for believing one thing about the world rather than its opposite, but *our sensory experience is all we have to go on*".[5]

---

[2]Or if not false, then groundless. The widely reported (and however implausible) case of the Duke of Devonshire, who dreamt that he was giving a speech in the House of Lords when he was in fact giving a speech in the House of Lords, shows that a belief may be true yet without the right sorts of reasons to support it.

[3]Stroud (1984), *passim*.

[4]Descartes puts the point in just this way in the Sixth Meditation: "The first [general reason for doubting] was that every sensory experience I ever thought I was having while awake I can also think of myself as sometimes having while asleep; and since I do not believe that what I seem to perceive in sleep comes from things located outside me, I did not see why I should be more inclined to believe this of what I think while awake" (Descartes 1984, p. 53).

[5]Stroud (1984, p. 32), our italics. Similarly, Williams (2001, p. 75), says: "To get to radical skepticism ['the thesis that our beliefs are completely unjustified'], the sceptic must not concede that his possibilities are remote. He must argue that they are as likely to be true as what we ordinarily believe. This is what he does. His point is that his bizarre stories about Evil Deceivers and brains-in-vats are just as likely to be true as our ordinary beliefs *given all the evidence we will ever have*. In the case of the external world, all the evidence we will ever have comes from our sensory experience; … in every case, he will claim, all the evidence we will ever have radically undermines what it would be true or even justifiable to believe" (his italics). It follows, Williams thinks, that we have no reason to believe that we are awake, given that we might be dreaming. This is a *non-sequitur*. It turns on the near-universal conflation of "evidence" with "justification".

But We Have Good Reasons to Believe that We Are Often Awake

But of course sensory experience is *not* "all we have to go on". In the absence of much evidence, we might still have good grounds for belief. Such grounds could very well include the fact that most of us are awake much more often than we are asleep, and that even when asleep, we do not always dream.[6] Which is to say that while the sensory-experience data do not provide much evidence that we are, at any given time, conscious (or dreaming), the posterior probability that we are conscious is much higher than the posterior probability that we are dreaming asleep, a function of the higher probability that physiological theory accords to the hypothesis that we are conscious. Hence we have good reason to believe, at any given time, that we are conscious (even on those occasions when we are, in fact, dreaming).[7] It is only if we run "justification" and "evidence" together, as Stroud in company with all of the other skeptics does, that we can conclude (given the possibility that we might at any moment be dreaming) that "we have no basis for believing one thing about the world rather than its opposite".

Our reason for believing that we are conscious rather than dreaming must be distinguished carefully from another, very different, criticism that is sometimes made of the argument from dreaming. This criticism is to the effect that a premise of Descartes' argument is that he remembers "other occasions when [he has] been tricked by exactly similar thoughts [to those he is now having while asleep]!" If Descartes remembers this, then it must be taken for a fact. But his knowledge of this fact is incompatible with the intended conclusion of the dreaming argument, that he knows nothing of the kind.

Stroud rightly points out that the argument does not depend on Descartes' knowledge of any facts. All that is necessary is that it be *possible* that he is asleep at any given time, and this "possibility" can be invoked quite independently of any presumed knowledge on his part that he has often dreamt that he was awake.

(Footnote 5 continued)

Granted that although there is little evidence properly so-called in the radical skeptical cases, we might still have very good reasons for belief.

[6]In fact, we are asleep roughly a third of the time, and dream roughly a fifth. While the role of sleep, still less that of dreams, is not yet fully understood, it is nonetheless clear that in the ordinary course of events, nutrition, among other bodily requirements, can (except when hooked to a feeding tube) be met only when we are awake. Not all of us can sleep, hence dream, all of the time (though some friends of ours make a very good pretense). The general contingencies of survival demand that we be more often awake. Indeed, horses and goats sleep only two to four hours a day, apparently to maximize their foraging time and minimize their vulnerability to predators. See Anch et al. (1988).

[7]In our view we are "tricked" into believing we are awake when in fact we are asleep, not so much by the fact that our dreaming experiences are "qualitatively indistinguishable" from our conscious experiences, for often they are not, as by the fact that we understandably tend, other things being equal, to believe that we are awake (even when we are not) as a heuristic rule.

It is necessary to be as clear as possible on this point so as to avoid misunderstanding, for it might otherwise appear that we have begged the question. If we take the presumptive fact that we are awake more often than not to justify our belief that we are awake, then, a critic might say, we have to justify that it *is* a fact. But this, of course, just pushes the question back to whatever grounds are cited as justifying the claim that we are awake more often than not. If we have no evidence for those grounds (as we claim), then how can they justify us? The regress continues, in which case we have not shown that the sceptic is mistaken.

It should be clear, first, that this sort of criticism trades on the very conflation between evidence and justification that it is our purpose to combat. That is, it simply assumes that if we have little evidence for a claim then it cannot possibly be justified.[8] But it does not follow from the claim, granted for the purposes of argument, that the likelihood of sequences of sense-experiences on the waking and dreaming hypothesis is virtually the same that we have no more reason to believe that we are awake than we are asleep.[9] The second key to the dissolution of the skeptical argument is to show that we do have more reason to believe that we are awake than asleep. On the Bayesian model of confirmation that we have adopted for reasons already given, this comes to showing that the waking hypothesis has a higher prior probability.

The Prior Probability that We Are Conscious Rather Than Dreaming Is Higher

Although all Bayesian probabilities are characterized in terms of belief and degrees thereof, and in this sense are "subjective", there is disagreement among Bayesians concerning the extent to which their determination is "objective" or "subjective". We have already commented on the issue briefly. It suffices for our purposes at the moment to make the following three points.

---

[8]The move is standard in the Cartesian secondary literature. Thomson (2000, p. 33), for example. Descartes' point "…is rather that we have no internal evidence or criteria which surely distinguishes dreaming and waking experience". From which it follows that "Any particular experience could be a dream", i.e., we would have no reason to believe of any particular experience that it was not a dream.

[9]It is explicit in Descartes, implicit in the arguments of many of those who think that the argument from dreaming is sound, that the reasons for our belief that we are awake must be "conclusive", i.e., cannot themselves be doubted. In the traditional vocabulary, whatever qualifies as a "reason" must be able to serve as a *foundation* for the rest of our knowledge. But as we indicated in Chap. 1, an important premise of this monograph is that what we believe on the basis of what we experience has no "foundation" in this sense. It is invariably uncertain. The task is not to evade this fact, but to provide a way of distinguishing between well- and poorly-grounded beliefs, even to the point of quantifying the degree to which they are well- and poorly-grounded. Whether it qualifies as "knowledge" or not, it is more than enough to support the claim that science provides us with very well-grounded beliefs about the physical world.

First, the way in which the disagreement is often settled is not open to us here (although it was in Chap. 6 when we discussed TB and PAP diagnostic cases). Wesley Salmon, for example, wisely says that

> The moral I would draw concerning prior probabilities is that they can be understood as our best estimates of the frequencies with which certain kinds of hypotheses succeed … If … one wants to construe them as personal probabilities, there is no harm in it, as long as we attribute to the subject who has them the aim of bringing to bear all his or her experience that is relevant to the success or failure of hypotheses similar to that being considered.[10]

We think there is something to this line of argument, and have already suggested that the relative frequency with which we are conscious as against dreaming can be considered a reason why we are justified in believing, other things being equal, that we are awake (even when we are asleep).

Still, the skeptic might very well object that to invoke the subject's "best estimates of the frequencies with which certain kinds of hypotheses succeed" *is*, in this case, to beg the question against her. For it is precisely one purpose of the argument from dreaming to open up the possibility that the waking hypothesis has never "succeeded", i.e., been justified by the subject's past "experience". The whole notion of "empirical success" has itself been undermined, and with it any talk of (objective) "frequencies".

In order to deal with this objection, we could argue along another line. Salmon points us in a usable direction. It is his more general position that

> the prior probabilities in Bayes Theorem can best be seen as embodying the kinds of plausibility judgments that scientists regularly make regarding the hypotheses with which they are concerned …. Plausibility arguments serve to enhance or diminish the probability of a given hypothesis prior to—that is, without reference to—the outcome of a particular observation or experiment.[11]

We have already used a "plausibility argument" to show why the greater relative frequency of our being awake than asleep is to be expected. It is among the requirements of human survival and therefore explains why we are more often awake. Salmon himself goes on to suggest various general criteria—formal, material, and pragmatic—that can be brought to bear in assessing the plausibility of particular hypotheses and thus of determining their prior probabilities.

In the case at hand, simplicity is perhaps the most important of these criteria. The waking hypothesis is, on the face of it, so much simpler than the dreaming hypothesis. One way in which to bring this out is to note how much machinery is needed to instantiate that beloved realization of the possibility that all of our experience is merely virtual, the classic science fiction movie *The Matrix*. We may be perpetually deceived, but it takes some real doing on the part of the deceiver! Think of all the hooking up that has to be done to bring us to believe that we are not brains in vats on the hypothesis that we are. Why not more simply believe that at

---

[10]Salmon (1990, p. 270).

[11]*Ibid*., p. 264.

least most of the time we are not deceived in thinking that we are awake? This consideration is reinforced by another: the Darwinian theory to which we appealed earlier to explain the ostensible fact that we are more often awake than asleep is itself the simplest of those currently available which have a scientific character, e.g., are testable in principle, and so on.[12] Put another way, we can have a priori reasons for distinguishing between "empirically equivalent" theories. Copernicus and some of his successors urged the superiority of his theory at least in part for such reasons.[13]

Thought-Experiments

Third, although Salmon does not mention this, plausibility arguments, in science and without, often take the form of thought-experiments. Consider the case that Galileo makes for his claim that objects fall at the same rate, regardless of their weight.[14] Assume, with Aristotle, that heavy bodies fall faster than light ones. If we were to attach a heavy object to a light one, then the compound object, being heavier, would have to fall faster than either of its components. But since the light object, falling more slowly, would act as a drag on it, the compound object would fall more slowly than the heavy object. Having derived this contradiction, we can infer that its Aristotelian premise is false. Insofar as our "data" in this case have to do with the visual observations we might make, it does little to support his view; indeed, Galileo's *opponents* sometimes appealed to them. It is clear that Galileo arrived at this law, not by way of sense-experience, but on the basis of a thought-experiment.[15] That is, he argued in an a priori way that objects fall [in a vacuum] at the same rate, which is to say in our terms that he assigned his generalization a very high-valued prior probability. Although the observational data for his claim vis-à-vis Aristotle's were negligible, it was nonetheless highly justified.

---

[12]In this case, as in many others, formal must be distinguished from causal simplicity. As Miller (1987, p. 247), reminds us, "by adding a variety of novel propositions…evolutionary theory reduces the formal simplicity of science. [But] an enormous gain in causal simplicity results. For a variety of regularities which ought, according to all rival frameworks, to have causes only have causes once the evolutionary principles are added".

[13]Although again, his argument is more persuasive on causal than on formal grounds. See Kuhn (1959, pp. 169–171). We are following the conventional wisdom, of course, in claiming that they are empirically equivalent, for in fact they are not.

[14]See Brown (1991) for an account of this and other apparently data-independent plausibility arguments.

[15]See Galileo (1974, pp. 66–67). In preparation for his thought-experiment, Galileo had already noted (p. 64) that "where we lack sensory observations, their place may be supplied by reasoning which is no less capable of understanding the change of solids by rarefaction and resolution than [the change of] tenuous and rare substances by condensation".

A very large literature has grown up around the concept of a thought-experiment, and many different analyses of it have been given. The only dimension which concerns us here has to do with whether or not the use of thought-experiments, at least on occasion, "transcends" experience", as the Galilean example suggests. John Norton, for example, argues that "in so far as they tell us about the world, thought experiments draw upon what we already know of it, either explicitly or tacitly; they then transform that knowledge by disguised argumentation".[16] From this it might seem to follow that the use of thought-experiments, drawing on our "knowledge of the world" to establish prior probabilities, once again begs the question against the sceptic.

It is enough to point out, however, that the argument from dreaming is itself a thought-experiment. If it "draws upon what we already know", then there is something incoherent about the conclusion that the skeptic draws from it.[17] If there is nothing incoherent about the skeptic's reasoning, then it does not draw upon what we already know. To put it briefly, the use of thought-experiments to determine prior probabilities in this particular context cannot beg any questions.

We must be clear about this point. To say that it is plausible that at least some of our experience is waking, on the basis of something like a thought-experiment, is not to deny that it is *possible* (in some sense of the word) that we are always asleep, that we are in a "Matrix" scenario, that we are brains in vats, and so on. Plausibility arguments do not establish *certainty*, still less *infallibility*. But that fact does not entail that we do not have good reason to believe that at least some of our experience is waking.[18] On our understanding of it, the dreaming argument is not the simple-minded challenge, "but can you be *sure* that any empirical claim you might make might not be false, given the possibility that you are dreaming?" Since this question is tantamount to the question, "can you be *sure* that you are not in a situation which has been set up in such a way that you cannot be sure that you are in

---

[16]Norton (2002, p. 44).

[17]A point apparently not lost on Descartes, for on his account dream images could only have been derived from prior waking experience (see the following footnote and, and for an earlier reference in the First Meditation—"it must surely be admitted that the visions which come in sleep are like paintings, which must have been fashioned in the likeness of things that are real"—Descartes (1984, p. 13). He has several reasons for moving rather quickly to the supposition of an Evil Demon; this is one of them.

[18]As part of his eventual response to skepticism concerning the existence of objects independent of our ideas of them, Descartes essays his own plausibility argument for assigning the "object hypothesis" a higher prior. On this hypothesis, "I can … easily understand that this is how imagination comes about … and since there is no other equally suitable way of explaining imagination that comes to mind, I can make a probable conjecture that the body exists". He goes on, of course, to say that "this is only a probability" and "not the basis for a necessary inference that somebody exists", but continuing interest in the argument from dreaming does not derive from his insistence on certainty, but rather from the more general claim that we can have *no reason* not to believe that we are dreaming. See Descartes (1984, p. 51).

that situation?" it verges on incoherency. The argument rests, rather, on the complex inference from the premise that there is no way to distinguish waking and dreaming experiences to the lemma that there is no evidence that we are awake to the theorem that we can have no good reasons for believing that we are awake. As we have tried to demonstrate in this monograph generally, and more particularly with respect to the present instance of it, this inference is faulty.[19]

The Problem of Under-Determination

The reader will have noted a pattern in many of the skeptical arguments we have discussed. They go from the under-determination of hypotheses by the evidence available for them—scientific measurements, black ravens, green emeralds, sense-experience—to paradoxical, eventually skeptical, conclusions—that hypotheses are not confirmed or disconfirmed selectively but only as part of a global constellation of theories, that white shoes as well as black ravens confirm the hypothesis that all ravens are black, that the hypothesis that all emeralds are grue is as well supported by the same evidence as the hypothesis that they are all green, that for all we know (or are capable of knowing) we live in a dream world. We have argued that all of these arguments (and there are many more that fit the same pattern) rest on a conflation of evidence with confirmation: since the paradoxical or skeptical hypothesis has as much "evidence" going for it as the conventional or commonly accepted hypothesis, it is equally well supported by the data and there is nothing to choose between them. What needs to be emphasized is, first, that data that fail to discriminate hypotheses do not constitute (in particular contexts) "evidence" for any of them, i.e., that "data" and "evidence" are not interchangeable notions, second, that it does not follow from the fact that there is strong evidence for a hypothesis as against one or more of its competitors that it is therefore highly confirmed, third, that it does not follow from the fact that a hypothesis is highly confirmed that there is strong evidence for it against its competitors. As a result, there is little need to invoke such desperate expedients as God's goodness, Descartes' own solution to the otherwise irremediable solipsism to which the argument from dreaming appears to condemn us, to rescue us.

---

[19]However deeply ingrained it is. J.L. Austin, among many others, thinks that the only way to avoid the conclusion is to deny the premise. In his view, there are all kinds of important differences between waking and sleeping experiences. Thus he writes in Austin (1962, p. 48): "I may have the experience … of dreaming that I am being presented to the Pope. Could it be seriously suggested that having this dream is 'qualitatively indistinguishable' from *actually being* presented to the Pope? Quite obviously not". End of discussion! Here as so often in philosophy and elsewhere, it is not the premise that is to blame but the inferential pattern, use of which is subsequently made. Our aim is to enlarge the usual kit-bag to include patterns of uncertain inference that are not typically combined.

References

Anch, M., Browman, C., Miller, M., & Walsh, J. (1988). *Sleep: a Scientific perspective*. Englewood Cliffs, NJ: Prentice-Hall.

Austin, J. L. (1962). *Sense and sensibilia*. New York: Oxford University Press.

Brown, J. (1991). *Laboratory of the mind: thought experiments in the natural sciences*. London: Routledge.

Descartes, (1984). *The philosophical writings of Descartes, translated by Cottingham, Stoothoof, and Murdoch* (Vol. II). Cambridge: Cambridge University Press.

Galileo, (1974). *Two new sciences, translated by S*. Drake. Madison: University of Wisconsin Press.

Kuhn, T. (1959). *The Copernican revolution*. New York: Vintage.

Miller, R. (1987). *Fact and method: explanation, confirmation, and reality in the natural sciences*. Princeton: Princeton University Press.

Norton, J. (2002). *Contemporary debates in the philosophy of science*. Oxford: Blackwell.

Salmon, W. (1990). Rationality and objectivity in science, or Tom Kuhn meets Tom Bayes. In Savage (Ed.), *Scientific theories*, volume XIV, *Minnesota studies in the philosophy of science*. Minneapolis: University of Minnesota Press.

Stroud, B. (1984). *The significance of philosophical scepticism*. Oxford: Oxford University Press.

Thomson, G. (2000). *On Descartes*. Belmont, CA: Wadsworth Publishing.

Williams, M. (2001). *Problems of knowledge*. Oxford: Oxford University Press.

Chapter 11
Concluding Reflections

Abstract Our object in this monograph has been to offer analyses of confirmation and evidence that will set the bar for what is to count as each and at the same time provide guidance for working scientists and statisticians. Philosophy does not sit in judgment on other disciplines nor can it dictate methodology. Instead, it forces reflection on the aims and methods of these disciplines in the hope that such reflection will lead to a critical testing of these aims and methods, in the same way that the methods themselves are used to test empirical hypotheses with certain aims in view. In the Appendix we discuss an application of the confirmation/evidence distinction to an important problem in current ecological research and in the process suggest ways of settling some outstanding problems at the intersection of statistics and the philosophy of science.

Keywords Curve-fitting problem · "True-model" assumption

The curve-fitting problem is well known.[1] It has to do with somehow reconciling two conflicting desiderata, *goodness-of-fit* and *simplicity*. On the one hand, the more closely a family of curves fits the available data, the more complex and unwieldy it will be, the larger the probability of making erroneous predictions with respect to future data. On the other hand, the simpler a family of curves (as measured, say, by the paucity of parameters) on the basis of which to predict future data, the greater the probability of being at some distance from the "true" family of curves. The problem is to avoid both over-fitting and over-simplifying, in an intuitively plausible way. This involves providing and justifying criteria on the basis of which a choice of a family of curves to represent the data available should be made. It raises a number of complex statistical issues and does not have an easy or generally-accepted solution.

Philosophers of science have their own version of the curve-fitting problem. It has to do with reconciling something like the same two desiderata, goodness-of-fit

---

[1] See Forster and Sober's classic paper on the subject (1994) followed by its evaluation in Howson and Urbach's standard reference work (2006), Bandyopadhyay et al. (1996) and Bandyopadhyay and Boik (1999).

© The Author(s) 2016
P.S. Bandyopadhyay et al., *Belief, Evidence, and Uncertainty*,
Philosophy of Science, DOI 10.1007/978-3-319-27772-1_11

with paradigms of classical and contemporary scientific practice and simplicity of analyses of this practice which are for this very reason both insightful and explanatory. Again the problem is to avoid both over-fitting and over-simplifying.

As we indicated in Chap. 4, there are a variety of critics who think that the problem has no solution. The closer we get to an accurate description of scientific practice, even over the past three or four hundred years, the more difficult it is to find any sort of usable pattern in it, certainly nothing on the basis of which we could give guidance to future scientists. If philosophers of science see a pattern, it is only by taking currently-accepted ways of proceeding and imposing them on the past. In the process, they do not simply idealize, but radically distort the data, understanding Kepler and Newton, for instance, in ways in which Kepler and Newton did not understand themselves.

There is something to this criticism, of course. But, it derives its force from three assumptions traditionally made by philosophers of science from which we have tried to distance ourselves in this monograph.

The first assumption is that there is *one* pattern which is at the same time both descriptive and explanatory. Thus we have rather monolithic accounts collapsing the concepts of evidence and confirmation and applying them across the board. This is not a naïve assumption. It is made by many scientists and statisticians. It is also rooted in the fact that the simpler the pattern, the more easily accessible and explanatory it is. What has in fact happened in the case of almost all of the major accounts of confirmation and evidence set out in the 20th century is that in the attempt to handle counter-examples, many of them fanciful, they have become more and more complex and arcane, adding new conditions as further equants and epicycles to their patterns, and in the process losing much of their value to provide insight. Simplicity has many virtues. The mistake is to limit the options to two: either hold on to the pattern, endlessly complicating it, or abandon the search for any pattern as a waste of time. On both options, the attempt to provide some sort of understanding of what is at stake when scientists carry out experiments to test hypotheses is given up.

Here as elsewhere we follow Kant's advice noted at the outset in Chap. 1 to examine carefully the shared premise in an on-going debate. Both sides to the particular debate about the possibility of a helpful philosophical analysis of confirmation/evidence agree that there is one pattern or there is none. On our view, to the contrary, there are at least two—confirmation, which we have characterized in a Bayesian way in terms of up-dating belief-probabilities in the light of new or re-considered data, and evidence, characterized in terms of likelihoods and their ratios. As we have tried to demonstrate, combining the two into some sort of unified account has led to a great deal of confusion. There is nothing wrong with unification: in philosophy as well as in science it has a great deal of explanatory power. But *pace* those who continue to look for "a theory of everything," the desire for it can sometimes lead us astray. We have argued that although the words "evidence" and "confirmation" are often used interchangeably, they in fact draw attention to two very different aspects of hypothesis testing. It won't do to substitute a great variety of accounts for one very complicated single account. But a small number of relatively simple accounts should be capable of providing genuine insight into at least some of the signal, but at bottom rather different, aspects of experimental science. In particular, we have tried to make some of Richard Royall's ideas concerning evidence known to a wider public in non-technical terms. As we said at the outset, our aim was not to add to existing paradigms, but to redeploy them and extend their applications. This is the aim of the first part of the monograph.

We have already dealt at length with the second assumption, that finding evidence for a theory at the same time provides us with good reasons for believing that the theory is true. Once again this "true-model" assumption opens the door to critics, most often of the attempt to gain insight into the structure of scientific inference, sometimes of science itself as an especially credible activity. The argument is that one "good reason" after another has been offered for theories that turned out to be false. Once burned, twice afraid. The historical evidence provides us with "good reasons" to believe that every theory that we now hold dear will one day be overturned, and perhaps that there is no such thing as truth. It won't do to argue either that *this time* we have the truth-obtaining methodology straight and we *now* have good reasons to believe that our theories are true or to rig the analysis in such a way that the possession of "evidence" rightly so call guarantees their truth. Better, we have claimed, to say that one theory is a better approximation to the truth than another, admit the in-principle possibility that all of our theories are false, and show how, in an objective way, some data constitute stronger evidence for a hypothesis than other data. A main, although not the only aim of the second part of the monograph, is to show how the "true-model" assumption operates in well-known conceptions of evidence other than ours and why it should be rejected.

Once the true model assumption has been rejected, and we assume instead that all of our models are but approximations, it might be wondered whether the concept of confirmation, and more narrowly the Bayesian conception of it, has much work to do. For however it is further analyzed, confirmation has to do with finding good reasons for believing that certain of our hypotheses are true. Moreover, the Bayesian conception that these "good reasons" are in some fundamental sense personal undermines both the objective and the communal character of science. On the one hand, objectivity is secured by agent-independent evidence. It does not rest on individual belief, however strongly held. On the other hand, the formation and testing of models is inter-personal and cumulative. Bayesianism identifies the learning "agent" as an individual, albeit ideally rational, person. But the truth is that the scientific community has learned much more than any individual scientist. This knowledge has accumulated over thousands of years through a complex web of transmission, colleague to colleague and teacher to student. It is this process that should be the fundamental object of a philosopher of science's attention.

We could put the point another way. We all learn from experience, some more than others, re-distributing probabilities over hypotheses. But such learning depends strongly on cultural background, physical state, and a host of other factors. It is not well suited to be a model of scientific investigation or large-scale scientific progress. Evidence is independent of these factors, it imposes norms, and it can accumulate over time. It is in these respects much better suited to form the basis of what might be called "public" epistemology.[2]

Obviously enough, the conduct and the growth of scientific knowledge cannot be entirely divorced from considerations of personal belief. Scientists are people, and create their research programs informed by these beliefs. One cannot very well describe scientific practice or lay an appropriate educational foundation for it, without also including a "personal" epistemology within the description. So, to provide a more comprehensive account of personal and public epistemology, the distinction between belief-related notions, in which confirmation and the true-model assumption play a role, and impersonal evidence is crucial. What is now needed is a deeper understanding of how bridges linking one to the other are best built.

The third assumption is closely linked to the second. It is that a philosophical analysis of the concepts of evidence and confirmation will set the bar for what is to count as each and provide guidance for working scientists and statisticians. Philosophy does not sit in judgment on other disciplines nor can it dictate methods. Instead, it forces reflection on the aims and methods of these disciplines in the hope that such reflection will lead to a critical testing of their aims and methods, in the same sort of way that the methods themselves are used to test empirical hypotheses. At the same time, we have tried to show how issues that might seem initially to be of interest only to philosophers of science have a much wider bearing. This is the point of the third part of our monograph: analyses of the concepts of evidence and confirmation, at least insofar as conducted in our way, are capable of not only resolving key paradoxes in the philosophy of science, but of shedding important light on the sorts of fundamental questions that all human beings raise about the world and our knowledge of it.

The hope is that our account has enough goodness-of-fit to scientific practice and enough simplicity and clarity to provide insight, and to put some of the issues mulled over by generations of philosophers in a new and larger context.

References

Bandyopadhyay, P., Boik, R., & Basu, P. (1996). The curve-fitting problem: A Bayesian approach. *Philosophy of Science, 63*(Supplement), 391–402.

Bandyopadhyay, P., & Boik, R. (1999). The curve-fitting problem: A Bayesian rejoinder. *Philosophy of Science, 66*, S390–S402.

Forster, M., & Sober, E. (1994). How to tell when simpler, more unified, or less ad hoc theories will provide more accurate predictions. *British Journal for the Philosophy of Science, 45*, 1–35.

Howson, C., & Urbach, P. (2006). *Scientific reasoning: The Bayesian approach* (3rd ed.). Chicago and LaSalle: Open Court Publishing.

Strevens, M. (2010). Reconsidering authority: Scientific expertise, bounded rationality, and epistemic backtracking. In T. S. Gendler & J. Hawthorne (Eds.), *Oxford studies in epsitemology* (Vol. 3). New York: Oxford University Press.

---

[2]See Strevens (2010).

Appendix
Projections in Model Space: Multi-model Inference Beyond Model Averaging

Mark L. Taper and José M. Ponciano

A reviewer of the manuscript challenged us to do two things. First, to move beyond simple likelihood ratio examples and show how evidential ideas are used in the practice of science. And, second to solve some deep problem in ecology using this framework. To answer the first challenge, we discuss information criteria differences as a natural extension of the likelihood ratio that overcomes many of the complexities of real data analysis. To answer the second more substantive challenge, we then extend the information criterion model comparison framework to much more effectively utilize the information in multiple models, and contrast this approach with model averaging, the currently dominant method of incorporating information from multiple models (Burnham and Anderson 2002). Model averaging is a confirmation-based-approach.

Because of limitations in both time and allowable word count, this will be a sketch of a solution.[1] We deeply appreciate the reviewer's challenge because the work it has forced us to do has been very rewarding.

Comparing Models Forms that are not Fully Specified

The evidential theory developed so far in this monograph has concerned fully specified models, that is, models with given parameter values. As such, they completely describe the probability distribution for data consistent with those models. Scientists, on the other hand, generally do not have that much foreknowledge. Instead, they commonly propose model forms, that is, models known up to the functional form of their elements, but without specific parameter values. Parameter values are estimated by fitting the model forms to data, and comparisons of these now fully-specified models made using the same data. As both estimated models do have likelihoods, it would seem straightforward to compare them with likelihood ratios, the measure which has been invoked in the body of this monograph to capture the evidential strength for one model against its rival.

---

[1] Technical statistical details will follow in other fora, beginning with a symposium at the Japanese Institute of Mathematical Statistical Mathematics in January of 2016.

Unfortunately, important problems do arise. The estimation of parameters creates a bias in the calculation of the probability of the data under the model and therefore it also creates a bias in the estimated likelihood of the model. This bias is related to the number of parameters in the model; the more parameters a model has, the greater the (positive) bias there is in the estimated likelihood. Carried to the extreme, if a model has as many parameters as there are data points, it can fit the data exactly, and its likelihood will become infinite. This despite having no predictive power.

Classically this problem has been handled (at least in the context of nested models) with the use of likelihood ratio tests. But, what about non-nested models? Constraining your models to be nested imposes a severe limitation on the ability of scientists to explore nature through models. Further, the likelihood ratio test is an error-statistical approach where the bias due to parameterization is accounted for in the critical value of the test. This book (Chap. 6) and Taper and Ponciano (2015) argue for the greater utility of an evidential approach compared to an error-statistical approach.

In 1973, Hirotogu Akaike wrote a brilliant paper developing the AIC, and thereby solving, in large part, the estimation bias problem of the likelihood ratio, allowing for a huge expansion in the scope of evidential analysis. Although Akaike referred to the AIC as "an Information Criterion," the AIC is universally termed the "Akaike Information Criterion."

Akaike with Tears

Technical accounts deriving Akaike's Information Criterion (AIC) exist in the literature (see for instance the general derivation of Burnham and Anderson 2002, Chap. 7), but few have attempted to clarify Akaike's (1973) paper, step by step. A notable exception is deLeeuw's (1992) introduction to Akaike (1973) Information Theory, which made it clear that more than a technical mathematical statistics paper, Akaike's seminal contribution was a paper about ideas: "…This is an 'ideas' paper, promoting a new approach to statistics, not a mathematics paper concerned with the detailed properties of a particular technique…"[2] deLeeuw then takes on the task of expunging the ideas from the technical probabilistic details and coming up with a unified account clarifying both the math and the ideas involved. His account is important because it makes evident that at the very heart of the derivation Akaike was using Pythagoras' theorem. It will be seen later that our contribution is to take this derivation one step further by using Pythagoras' theorem

---

[2]If this is their initial encounter with information criteria, we suggest that readers first familiarize themselves with a gentler introduction such as Malcolm Forster and Elliott Sober's "How to Tell When Simple, More Unified, or Less Ad Hoc Theories Will Provide More Accurate Predictions," which explains, develops, and applies Aikaike's ideas to central problems in the philosophy of science. *British Journal for the Philosophy of Science* 45: 1–35, 1994. A comparison between different standard information criteria can be found in Bandyopadhyay and Brittan (2002).

again. In what follows we will set the stage to explain our contribution using Akaike (1973, 1974) and deLeeuw (1992). Given this monograph's context, our account will focus more on the ideas than on the technical, measure theoretic details for the sake of readability and also because this approach will allow us to shift directly to the core of our contribution.

A central objective in scientific practice is trying to come up with some measure of comparison between an approximating model and the generating model. Following Akaike, we shall be concerned for the time being with the parametric situation where the probability densities are specified by a set of parameters $\theta = (\theta_1, \theta_2, \ldots, \theta_L)'$ in the form $f(x; \theta)$. The true, generating model will be specified by setting $\theta = \theta_0$ in the density f. Setting aside the fact that truth is unknown, under this setting the comparison between a general model and the true model can be done, as in the rest of the monograph, via the likelihood ratio $\tau(x, \theta, \theta_0) = \frac{f(x;\theta)}{f(x;\theta_0)}$ without loss of efficiency. This well-known statistical fact suggests using some discrimination function $\Phi(\tau(x, \theta, \theta_0))$ of the likelihood ratio between θ and the true model θ_0. The data, x, are random and so the average discrimination over all possible data would better represent the distance between a model and the truth. Such an average would then be written as

$$\mathcal{D}(\theta, \theta_0; \Phi) = \int f(x; \theta_0)\Phi(\tau(x, \theta, \theta_0))dx = \mathbb{E}_x[\Phi(\tau(x, \theta, \theta_0))],$$

where the expectation is over the sampled stochastic process of interest X. Akaike then suggested study of the sensitivity of this quantity to the deviation of θ from θ_0. Two questions of interest immediately arise: can such an average discrimination be minimized and if so, can its minimization be estimated from realized observations of the process?

To get at this quantity, Akaike thought of expanding it via a Taylor series around θ_0 and keeping a second order approximation, which we write here for a univariate θ:

$$\mathcal{D}(\theta, \theta_0; \Phi) \approx \mathcal{D}(\theta_0, \theta_0; \Phi) + (\theta - \theta_0)\frac{\partial \mathcal{D}(\theta, \theta_0; \Phi)}{\partial \theta}\bigg|_{\theta=\theta_0} + \frac{(\theta - \theta_0)^2}{2!}\frac{\partial^2 \mathcal{D}(\theta, \theta_0; \Phi)}{\partial \theta^2}\bigg|_{\theta=\theta_0} + \cdots$$

To write this approximation explicitly, note that $\tau(x, \theta, \theta_0)|_{\theta=\theta_0} = 1$. Also, note that since f is a probability density $\int f(x; \theta)dx = 1$, which together with the regularity conditions that allow differentiation under the integral sign results in $\int \frac{\partial f(x;\theta)}{\partial \theta}dx = \int \frac{\partial^2 f(x;\theta)}{\partial \theta^2}dx = 0$. Then, $\frac{\partial \mathcal{D}(\theta,\theta_0;\Phi)}{\partial \theta}\bigg|_{\theta=\theta_0} = 0$ and

$$
\begin{aligned}
\frac{\partial^2 \mathcal{D}(\theta, \theta_0; \Phi)}{\partial \theta^2}\bigg|_{\theta=\theta_0}
&= \int \frac{\partial}{\partial \theta}\left(\frac{\partial \Phi(\tau)}{\partial \tau}\frac{\partial \tau}{\partial \theta}\right) f(x; \theta_0) dx \bigg|_{\theta=\theta_0} \\
&= \int \frac{\partial^2 \Phi(\tau)}{\partial \tau^2}\left(\frac{\partial \tau}{\partial \theta}\right)^2 f(x; \theta_0) dx \bigg|_{\theta=\theta_0} + \int \frac{\partial^2 \tau}{\partial \theta^2}\frac{\partial \Phi(\tau)}{\partial \tau} f(x; \theta_0) dx \bigg|_{\theta=\theta_0} \\
&= \Phi''(1) \int \left(\frac{1}{f(x; \theta_0)}\frac{\partial f(x; \theta)}{\partial \theta}\right)^2 f(x; \theta_0) dx \bigg|_{\theta=\theta_0} \\
&= \Phi''(1) \int \left(\frac{\partial \log f(x; \theta)}{\partial \theta}\right)^2 f(x; \theta) dx \bigg|_{\theta=\theta_0} = \Phi''(1)\mathcal{I}(\theta_0),
\end{aligned}
$$

where $\mathcal{I}(\theta_0)$ is Fisher's information. In going from the first line to the second line, a combination of the product rule and chain rule is employed. In going from the second line to the third, it is noted that the results given immediately above indicate that the right hand integral is 0. This preliminary result is non-trivial because it demonstrates that the resulting approximation of the average discrimination function

$$
\mathcal{D}(\theta, \theta_0) \approx \Phi(1) + \frac{1}{2}\Phi''(1)(\theta - \theta_0)^2 \mathcal{I}(\theta_0)
$$

is directly scaled by the theoretical variance of the Maximum Likelihood Estimator (MLE). Next, Akaike proposed using the functional form $\Phi(t) = -2\log(t)$ so that $\mathcal{D}$ behaves like a distance, i.e., is always non-negative and $\mathcal{D}(\theta_0, \theta_0) = \Phi(1) = 0$. The factor of 2 is conventional. Conveniently, the approximation, from here on denoted as $\mathcal{W}(\theta, \theta_0)$ then becomes $\mathcal{D}(\theta, \theta_0) \approx \mathcal{W}(\theta, \theta_0) = (\theta - \theta_0)^2 \mathcal{I}(\theta_0)$. It is straightforward to show that in the multivariate case, the approximation is written as the quadratic form $\mathcal{W}(\theta, \theta_0) = (\theta - \theta_0)'\mathcal{I}(\theta_0)(\theta - \theta_0)$, where $\mathcal{I}(\theta_0)$ is Fisher's information matrix. On the other hand, inserting Akaike's choice of a functional form into the original definition of the average discrepancy gives

$$
\begin{aligned}
\mathcal{D}(\theta, \theta_0) &= -2 \int f(x; \theta_0) \log\left(\frac{f(x; \theta)}{f(x; \theta_0)}\right) dx \\
&= -2\mathbb{E}_X\left[\log \frac{f(X; \theta)}{f(X; \theta_0)}\right] \\
&= -2[\mathbb{E}_X(\log f(X; \theta)) - \mathbb{E}_X(\log f(X; \theta_0))] \\
&= 2\mathbb{E}_X(\log f(X; \theta_0)) - 2\mathbb{E}_X(\log f(X; \theta)).
\end{aligned}
$$

This form of the average discrimination function is known as the negentropy, or the Kullback–Leibler (KL) divergence. So from the start, Akaike was able to make two crucial connections between his choice measure of discrepancy between the true generating model and an approximating model. One, directly bringing the theory of ML estimation into the scaling of such discrepancy, and the other, linking these concepts with a wealth of results in Information Theory. Thus, it was natural for Akaike to call $\mathcal{D}(\hat{\theta}, \theta_0)$ the probabilistic entropy. If $X_1 = x_1, X_2 = x_2, \ldots, X_n = x_n$ observations from the process X are available, then using the law of large numbers the KL divergence (or probabilistic entropy) could be estimated consistently with the average likelihood ratio

$$\widehat{\mathcal{D}}_n(\hat{\theta}, \theta_0) = -2 \times \frac{1}{n} \sum_{i=1}^{n} \log \frac{f(x_i; \hat{\theta})}{f(x_i; \theta_0)},$$

where $\hat{\theta}$ is the MLE. In reality, one cannot compute this likelihood ratio because the true model θ_0 in the denominator is unknown. However, because for every data point x_i the denominator in this average log-likelihood ratio is the same constant, Akaike pointed out that even if truth is unknown, we do know that maximizing the (log-) likelihood also minimizes the KL divergence between the estimated density and the true density. This is why Akaike called his contribution an "extension of the principle of maximum likelihood". Not content with this result, and in a remarkable display of the reaches of frequentist thinking, Akaike pointed out that because multiple real-izations of the array of data points $X_1, X_2, \ldots, X_n$ yield multiple estimates of θ_0, one should in fact think of the average discrepancy as a random variable, where the randomness is with respect to the probability distribution of the MLE $\hat{\theta}$. Therefore one may in fact be able to minimize the KL divergence between the true generating model and the approximating model by minimizing the average of $\mathcal{D}(\hat{\theta}, \theta_0)$—averaged over the distribution of $\hat{\theta}$. The problem of minimization of the KL divergence then becomes a problem of approximation of an average, something that statisticians are (suppos-edly) good at. Let $\mathcal{R}(\theta_0) = \mathbb{E}_{\hat{\theta}}\left[\mathcal{D}(\hat{\theta}, \theta_0)\right]$ denote our target average. Substituting the probabilistic entropy by its definition using the expectations over the process we get

$$\mathcal{R}(\theta_0) = \mathbb{E}_{\hat{\theta}}\mathcal{D}(\hat{\theta}, \theta_0) = 2\mathbb{E}_{\hat{\theta}}\left[\mathbb{E}_X(\log f(X; \theta_0)) - \mathbb{E}_X\left(\log f(X; \hat{\theta})|\hat{\theta}\right)\right]$$
$$= 2\mathbb{E}_X(\log f(X; \theta_0)) - 2\mathbb{E}_{\hat{\theta}}\left[\mathbb{E}_X\left(\log f(X; \hat{\theta})|\hat{\theta}\right)\right].$$

The first term in this expression is an unknown constant whereas the second term is a double expectation. Instead of working directly with these expectations, Akaike thought of substituting for the probabilistic entropy $\mathcal{D}(\hat{\theta}, \theta_0)$ by its quadratic approximation $\mathcal{W}(\hat{\theta}, \theta_0)$ via a Taylor series expansion and a very creative and useful way to re-write this expression. Akaike noted that the quadratic form

$$\mathcal{W}(\theta, \theta_0) = (\hat{\theta} - \theta_0)' \mathcal{I}(\theta_0)(\hat{\theta} - \theta_0)$$

used to approximate $\widehat{\mathcal{D}}_n(\hat{\theta}, \theta_0)$ can be seen (as any quadratic form involving a positive definite matrix and a fixed point) as the squared statistical distance between $\hat{\theta}$ and θ_0. This is the square of a statistical distance because proximity between

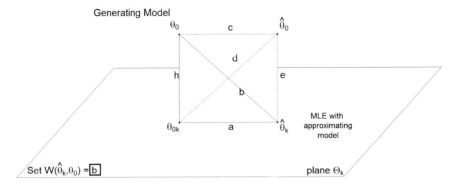

Fig. 1 The geometry of Akaike information criteria. θ_0 is the generating model. θ_{0k} is the closest projection of the generating model onto the plane, Θ_k, defined by the approximating models. $\hat{\theta}_k$ is the maximum likelihood estimate of an approximating model, which can be though of as the orthogonal projection of the estimator of θ_0 by an approximating model with the same dimension as the generating model, that lies in the same surface as θ_0 (the upper right vertex of (h,a,e,c)). The diagonal b is the discrepancy between the MLE of an approximating model and the generating model. The vertical h is the discrepancy between the generating model and its best possible projection to the plane. Edge h is fixed but edges a, e and c are random. Akaike showed that $d^2 = h^2 + c^2 - q$, where q is the inner product between c and h, and is much smaller than the other (quadratic) terms. Akaike then used the expected value of the approximation $d^2 = h^2 + c^2$ to derive his statistic. Depicted is such approximate configuration.

points is weighted by the dispersion of the points in the multivariate space, which is in turn proportional to the eigenvalues of the positive definite matrix $\mathcal{I}(\theta_0)$ (see plots explaining the geometric interpretation of quadratic forms in Johnson and Wichern 2002, Chaps. 1 and 2). Expressing the average discrepancy as the squared of a distance was a crucial step in Akaike's derivation because it opened the door for its decomposition using Pythagoras' theorem. By doing such decomposition, one can immediately visualize through a simple sketch the ideas in his proof (see Fig. 1). The first vertex of Akaike's right-angle's triangle is the truth θ_0 of dimension L (unknown). The second vertex is the estimator $\hat{\theta}$ of dimension $k \leq L$, as it comes from an approximating model. We will denote this estimator as $\hat{\theta}_k$ to emphasize its smaller dimension. The third vertex is θ_{0k} which is the orthogonal projection of the truth in the plane where all estimators of dimension k lie, that we will denote Θ_k. A fourth point crucial to derive the AIC is given by the estimator of θ_0 from the data using an approximating model with the same dimension as θ_0. To distinguish it from $\hat{\theta}_k$ we will denote it as $\hat{\theta}_0$. This estimator can be thought of lying in the same plane as θ_0. In the sketch in Fig. 1, we have labeled all the edges with a lowercase letter. To make the derivation as simple as possible, we will do the algebraic manipulations with these letters. In so doing, we run the unavoidable risk of trivializing one of the greatest findings of modern statistical science, all for the sake of transmitting the main idea behind the proof. The reader, however, should be well aware that these edges (lower case letters) denote, by necessity, random variables and that in the real derivation, more complex arguments including limits in probability and fundamental probability facts are needed.

Given the law of large numbers approximation of the average discrepancy using the average log-likelihood ratio, Akaike's initial idea was to use, as estimate of $\mathcal{W}(\hat{\theta}, \theta_0), \widehat{\mathcal{D}}_n(\hat{\theta}_k, \hat{\theta}_0) = -2 \times \frac{1}{n} \sum_{i=1}^{n} \log \frac{f(x_i; \hat{\theta}_k)}{f(x_i; \hat{\theta}_0)}$ ($=e^2$ in the sketch below) and let $n \to \infty$. However, the Pythagorean decomposition illustrated below shows that the estimated discrepancy $\widehat{\mathcal{D}}_n(\hat{\theta}_k, \hat{\theta}_0)$ will be a biased estimate of the target discrepancy because of the substitution of the ML estimators. One point in favor of the usage of $\widehat{\mathcal{D}}_n(\hat{\theta}_k, \hat{\theta}_0)$, however, is that ML theory tells us that $n\widehat{\mathcal{D}}_n(\hat{\theta}_k, \hat{\theta}_0)$ is chi-square distributed with degrees of freedom $L - k$. With this result at hand, and using simple geometry, Akaike sought to re-write the Pythagorean decomposition using $\widehat{\mathcal{D}}_n(\hat{\theta}_k, \hat{\theta}_0)$. The last piece of the puzzle needed to be able to do that was to demonstrate via convergence in probability calculations that the edge $a = (\hat{\theta}_k, \theta_{0k})$ was the stochastic projection of the edge $c = (\hat{\theta}_0, \theta_0)$ in the Θ_k plane. Below is the sketch aforementioned:

In simple terms, the objective is to solve for the edge length b using what we can estimate (e^2 through the log-likelihood ratio $n\widehat{\mathcal{D}}_n(\hat{\theta}_k, \hat{\theta}_0)$). Using Pythagoras' theorem we get that

$$b^2 = h^2 + a^2. \tag{1}$$

Note also that $d^2 = e^2 + a^2$ so that

$$e^2 = d^2 - a^2. \tag{2}$$

However, arguing that the third term of the squared distance $d^2 = c^2 + h^2 - 2c.h.\cos\phi$ remained insignificant compared with the other squared terms, Akaike re-wrote it as $d^2 \approx c^2 + h^2$, which upon substituting into (2) gives

$$e^2 = h^2 + c^2 - a^2. \tag{3}$$

Now, doing (1–3) gives $b^2 - e^2 = h^2 + a^2 - h^2 - c^2 + a^2$. Hence, it follows that

$$b^2 = e^2 + 2a^2 - c^2. \tag{4}$$

Expressing the square distances in Eqs. (1–3), expanding them using Taylor Series expansions, estimating Fisher's Information in each case with the observed information and using convergence in probability results, Akaike was able to show that

$$nc^2 - na^2 \sim \chi^2_{L-k},$$

and that

$$na^2 \sim \chi_k^2,$$

so that Eq. (4) multiplied by n can be re-written as

$$nb^2 = n\mathcal{W}(\hat{\theta}_k, \theta_0) \approx \underbrace{n\mathcal{D}_n(\hat{\theta}_k, \hat{\theta}_0)}_{=\log-\text{likelihood ratio}} + \underbrace{na^2}_{\sim \chi_k^2} - \underbrace{n(c^2 - a^2)}_{\sim \chi_{L-k}^2}.$$

The double expectation from the original average discrepancy definition is then implemented by simply replacing the chi-squares by their expectations, which immediately gives

$$n\mathbb{E}_{\hat{\theta}_k}\left[\mathcal{W}(\hat{\theta}_k, \theta_0)\right] \approx n\mathcal{D}_n(\hat{\theta}_k, \hat{\theta}_0) + 2k - L, \text{ or}$$

$$\mathbb{E}_{\hat{\theta}_k}\left[\mathcal{W}(\hat{\theta}_k, \theta_0)\right] \approx \frac{-2}{n}\sum_{i=1}^{n}\log f(x_i; \hat{\theta}_k) + \frac{2k}{n} - \frac{L}{n} + \frac{2}{n}\sum_{i=1}^{n}\log f(x_i; \hat{\theta}_0). \tag{5}$$

Recall that what Eq. (5) is approximating is in fact

$$\mathcal{R}(\theta_0) = \mathbb{E}_{\hat{\theta}}\mathcal{D}(\hat{\theta}_k, \theta_0) = -2\mathbb{E}_{\hat{\theta}}\left[\mathbb{E}_X\left(\log f(X; \hat{\theta}_k)|\hat{\theta}_k\right)\right] + 2\mathbb{E}_X(\log f(X; \theta_0)), \tag{6}$$

which is the expected value (with respect to $\hat{\theta}_k$) of

$$-2\int f(x; \theta_0)\log\frac{f(x; \hat{\theta}_k)}{f(x; \theta_0)}dx = -2\int f(x; \theta_0)\log f(x; \hat{\theta}_k)dx + 2\int f(x; \theta_0)\log f(x; \theta)dx. \tag{7}$$

Using Eq. (5) where the first two terms are known and the next two terms include the unknown dimension, and the law of large numbers approximation of the first integral, Akaike concluded that an unbiased estimation of the expected value over the distribution of $\hat{\theta}_k$ of the first integral would be given by the average of the first two terms in Eq. (5).

The first term in Eq. (5) is $(-2/n)$ times the log likelihood with the approximating model. The last two terms cannot be known, but because upon comparing various models they will remain the same can be ignored in practice. Because n also remains the same across models, in order to compare an array of models one only has to compute $AIC = -2\sum_{i=1}^{n}\log f(x_i; \hat{\theta}_k) + 2k$ and choose the model with the lowest score as the one with the smallest discrepancy to the generating model. The logic can be graphically represented by Fig. 2 (drawn from Burnham et al. 2011)

In the popular literature (e.g. Burnham and Anderson 2002, p. 61, or Burnham et al. 2011) it is often asserted that the $-AIC/2$ is an estimator of $\mathbb{E}_{\hat{\theta}}\left[\mathbb{E}_X\left(\log f(X; \hat{\theta}_k)|\hat{\theta}_k\right)\right]$.

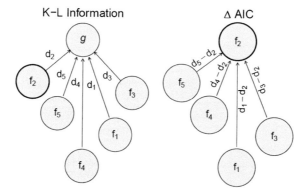

Fig. 2 The generating model is indicated by g, and the ith approximating models by f$_i$. Kullback–Leibler information discrepancies (d$_i$) are shown on the *left* as the distance between approximating models and the generating model. The Δ AIC shown on the *right* measure the distance from approximating models to the best approximating model. All distances are on the information scale

It is not, as Akaike (1974) states, the estimator of this quantity is –AIC/2n. For the qualitative comparison of models, this distinction makes no difference, but factoring the sample size (n) into the AIC allows a comparer of models to assess not only which model appears best, but what is the strength of evidence for that statement.

The Problem of Multiple Models

A model-centric view of science coupled with a disavowal of the absolute truth of any model pushes the scientist to the use of many models. Once this stance is taken, the question of how to use multiple models in inference naturally arises. Inference by the best model is not adequate as many models may be indistinguishable on the basis of pairwise comparison (see Chap. 2).

Currently, the dominant method for incorporating information from multiple models is model averaging. This comes in several flavors. In all cases model averaging is inherently, and generally explicitly, a Bayesian approach. Most common in ecology is averaging model parameter estimates or model predictions using Akaike weights. The Akaike weight for the ith model is given as:

$$w_i = \frac{\exp(-\Delta_i/2)}{\sum\limits_{r=1}^{R} \exp(-\Delta_r/2)},$$

where Δ_i is the difference between a model's AIC value and the lowest AIC value from the model set of R models indexed by r. Although it is not always pointed out, w_i is a posterior probability based on subjective priors of the form

$$q_i = C \cdot \exp\left(\frac{1}{2}k_i \log(n) - k_i\right)$$

where q_i is the prior for model i, C is a normalization constant, k_i is the number of parameters in the model, and n is the number of observations. The use of this prior makes model averaging a confirmation approach.

Two difficulties with model averaging for an evidentialist are: (1) the weights are based on beliefs, and are thus counter to an evidential approach. And (2) as a practical matter, model averaging does not take into account model redundancy. The more effort put into building models in a region of model space, the more heavily that region gets weighted in the average. We propose the alternative of estimating the properties of the best projection of truth, or a generating model, to the hyper-plane containing the model set. This mathematical development extends Akaike's insight by using the known KL distances among models as a scaffolding to aid in the estimation of the location of the generating model.

For convenience, we follow Akaike's (1974) notation and denote $Sgf = \int f(x; \theta_0) \log f(x; \hat{\theta}_k) dx$ and $Sgg = \int f(x; \theta_0) \log f(x; \theta) dx$, where the g refers to the 'generating' model and the f to the approximating model. Akaike's observation is then written as:

$$\widehat{Sgf} = \frac{1}{n} \sum_{i=1}^{n} \log f(x_i; \hat{\theta}_k) - \frac{k}{n} = -\frac{AIC}{2n}. \tag{8}$$

Accordingly, the KL divergence between a generating model g and an approximating model f can simply be written as $KL(g,f) = Sgg - Sgf$. From now on we will stick to this short-hand notation. One last detail that we have not mentioned so far is the fact that Akaike's approximation works provided $\hat{\theta}_k$ is close to θ_0 for any k. In fact, this is precisely why the Pythagorean decomposition works. The staggering and successful use of the AIC in the scientific literature shows that such approximation is in many cases reliable. Under the same considerations, we now extend these ideas to the case where we want to draw inferences from the spatial configuration of $f_1, f_2 \ldots$ approximating models to the generating model g.

The fundamental idea of our contribution is to use the architecture of model space to try to estimate the projection of truth onto a (hyper)plane where all the approximating models lie. Having estimated the location of truth, even without estimating it per se would anchor the AIC statistics in a measure of overall goodness of fit, as well as provide invaluable insights into the appropriateness of model averaging. The intuition of the feasibility of such a task comes from the realization that approximating models have similarities and dissimilarities. A modeler is drawn naturally to speak of the space of models. All that remains is to realize that that language is not metaphor, but fact. KL divergences can be calculated between any distributions and are not restricted to between generating processes and approximating models. A set of models has an internal geometrical relationship which constrains it and therefore has information about the relationship of approximating models and the generating process.

Computational advances have rendered straightforward algorithmic steps that while conceptually feasible would have been computationally intractable at the time that Akaike was developing the AIC. First, it is now easy to calculate the KL divergence between any two models. For instance, for the Normal distribution, the KL discrepancy can be computed exactly using the package gaussDiff in the statistical software R. Other packages will estimate the KL divergences of arbitrary distributions. Thus for a large set of approximating models, a matrix of estimated KL divergences among the set of models can be constructed. Second, parallel processing has tamed the computer intensive Non-Metric Multidimensional (NMDS) scaling algorithm which can take an estimated matrix of KL divergences and estimate the best Euclidean representation of model space in a (hyper)plane with coordinates $(y_1, y_2, \ldots)$. Nothing in our development restricts model space to be restricted to $\mathbb{R}^2$. To emphasize this we speak of a (hyper)plane, but to have any hope of visualizing we stay in $\mathbb{R}^2$ for this paper.

Suppose then that one can place the approximating models $f_1, f_2, \ldots$, on a Euclidean plane, as in the sketch below. For simplicity we have placed only two models in the sketch. Our derivation is not constrained to their particular configuration in the plane, relative to the generating model (truth), as the Fig. 3a, b show. Define m with coordinates $(y_1^{\star}, y_2^{\star})$ as the projection of the generating model (truth) in the Euclidean plane of models. This projection is separated by the length h to the generating model. Define $d(f_i, m)$ as the distance in the hyper(plane) of model i from m. Of course, the edges and nodes in this plane are random variables, associated with a sampling error. But, for the sake of simplicity and just as we did above to explain Akaike's derivation of the AIC, we conceive them for the time being as simple fixed nodes and edges.

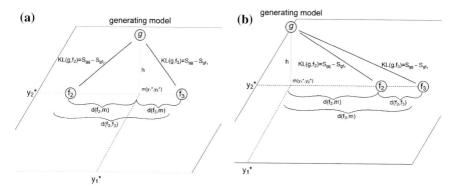

Fig. 3 The geometry of model space. f_2 and f_3 are approximating models residing in a (hyper)plane. g is the generating model. m is the projection of g onto the (hyper)plane. d(.,.) are distances between models in the plane. $d(f_2, f_3) \approx KL(f_2, f_3)$ with deviations due to the dimension reduction in NMDS and non-Euclidian behavior of KL divergences. As KL divergences decrease, they become increasingly Euclidean. Panel **a** shows a projection when m is within the convex hull of the approximating models, and Panel **b** shows a projection when m is outside of the convex hull

Then, using Pythagoras and thinking of the KL divergences as squared distances, the following equations have to hold simultaneously:

$$\begin{cases} KL(g,f_1) = d(f_1,m)^2 + h_1^2 \\ KL(g,f_2) = d(f_2,m)^2 + h_2^2 \\ \quad\vdots \end{cases}$$

where necessarily $h_1 = h_2 = h_i = \cdots = h$. In practice, one can decompose the KL divergence into an estimable component, Sgf_i and a fixed unknown component Sgg. Given that the Sgf_i are estimable as in Eq. (8), one can re-write the above system of equations including the unknown constants $Sgg, y_1^\star, y_2^\star$ as follows:

$$\begin{cases} Sgg - \widehat{Sgf_1} - d(f_1, m(y_1^\star, y_2^\star))^2 = h_1^2, \\ Sgg - \widehat{Sgf_2} - d(f_2, m(y_1^\star, y_2^\star))^2 = h_2^2, \\ \quad\vdots \end{cases} \tag{9}$$

Then, operationally, in order to estimate the location of the orthogonal projection of the generating model in the plane of approximating models, one can easily program the system of Eq. (9) into an objective function that, for a given set of values of the unknown parameters $Sgg, y_1^\star, y_2^\star$, computes the left hand sides of Eq. (9) and returns the sum of the squared differences between all the h_i^2. Then, a simple minimization of this sum of squared differences leads to an optimization of the unknown quantities (Fig. 4).

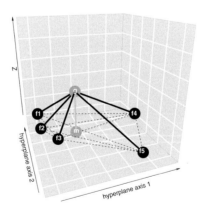

Fig. 4 The models of Fig. 2 visualized by our new methodology. As before, g is the generating model and {f1,...,f5}, are the approximating models. The *dashed lines* are KL distances between approximating models, which can be calculated. The *solid black lines* are the KL distances from approximating models to the generating model, which now can be estimated. The model labeled m is the projection of the generating model to the plane of the approximating models. The *solid gray line* shows h, the discrepancy between the generating model and its best approximation in the NMDS plane

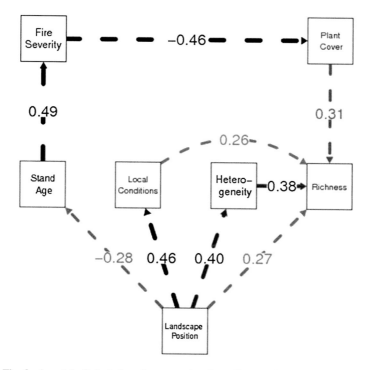

Fig. 5 The final model of plant diversity generation from Grace and Keely (2006) simplified by ignoring measurement error. *Arrows* indicate causal influences. The standardized coefficients are indicated by path labels and by path widths

We demonstrate this approach with a simulation based on the published ecological work of Grace and Keely (2006). Analyzing community composition data at 90 sites over 5 years, they studied the generation of plant community diversity after wildfire using structural equation models. Structural equation modeling is a powerful suite of methods facilitating the incorporation of causal hypotheses and general theoretical constructs directly into a formal statistical analysis (Grace and Bollen 2006, 2008; Grace 2008; Grace et al. 2010). The final model that Grace and Keely arrived at is shown in Fig. 5.

The figure should be read to mean that species richness is directly influenced by heterogeneity, local abiotic conditions, and plant cover. Heterogeneity and local abiotic conditions are themselves both directly influenced by landscape position, while plant cover is influenced by fire severity, which is influenced by stand age, which is itself influenced by landscape position. Numbers on the arrows are path coefficients and represent the strength of influence.

Our purpose in presenting this model is not to critique it or the model identification process by which it was found, but to use it as a reasonably realistic biological scenario from which to simulate. In short, we play god using this as a known true generating process. We consider in this analysis 41 models of varying complexity fitted to the simulated data. They cover a spectrum from underfitted to overfitted.

Fig. 6 NMDS space of 41
near approximating modes.
The true projection, **M**, of the
generating model to the
NMDS plane. The estimated
location of the projection, **m**,
and the location, **a**, of the
model average

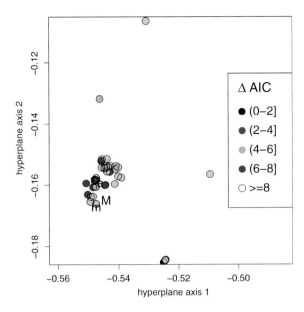

We calculated the NMDS 2-dimensional model space as described above. The
stress for this NMDS is extremely low (0.006 %) indicating the model space fits
almost perfectly into an $\mathbb{R}^2$ plane. We have plotted the fitted models in this space,
grey-scale coded by the AIC categories. We have also plotted in Fig. 6 the location
of our methods estimated projection of the generating model to the NMDS plane,
the model averaged location using Akaike weights, and the true projection of the
generating model to the NMDS plane (we know this because we are acting as God).
We can see in Fig. 6 that the estimated projection is slightly closer to the true
projection than is the model-averaged location.

In Fig. 7 we plot the effect on the estimated projection and model average of
deleting models from consideration. We sequentially delete the left-most model
remaining in the set, recalculating locations with each deletion. We see that the
model-averaged location shifts systematically rightward with deletion, and that the
location of the estimated projection is in this example more stable than the model
averaged location. It remains in the vicinity of its original estimate even after all
models in the vicinity of that location have been removed from consideration. If we
delete from the right, the model average moves systematically leftward. The model
projection location is, in this sequence, less stable than under deletion from the left.
These deletion exercises highlight several interesting facts about the two types of
location estimates that are implicit in the mathematics, but easily overlooked. First,
the model average is constrained to lie within the convex hull[3] of the approximating
model set. If you shift the model set, you will shift the average. Second, the

---

[3]The convex hull of a set of points in a plane is easily visualized as the space contained by a rubber
band placed around all of the points.

Fig. 7 Trajectories of the
model prediction and the
model average under deletion
of 1–30 models

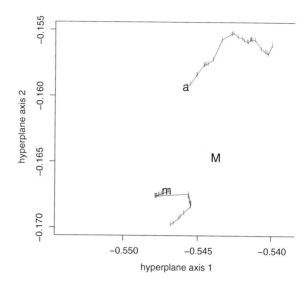

estimated generating model projection as a projection can lie outside of the convex hull. Third, because of the geometrical nature of the projection estimate, distant models can contribute information to the location of the best projection. This is the difference between rightward and leftward deletion. There are several models with high influence on the right hand side of the plot which are retained until the end in rightward deletion, but removed early in leftward deletion.

Unlike model averaging, the model projection methodology also produces estimates of two more quantities. The *Sgg*, the neg-selfentropy of the generating process, is estimated as −9.881. As God, we know that the true value is −9.877. These two agree to three significant figures. Also estimated is the distance of the generating process from the (hyper)plane of the NMDS model space. This is very important, because if the generating process is far from the (hyper)plane then any property estimate based on information from the model set should be suspect. The estimate for this discrepancy is 0.00018, indicating that is very close the (hyper)-plane. The true discrepancy is 5.8 e−08.

Discussion

This is just a brief sketch of an approach. Much statistical development, extension and validation are needed. These will be reported in other venues. Topics for development include:

Response surface methods to predict the properties of models near the generating model. The model average does weight models with low AIC more heavily than models with higher AIC, but does not take into consideration the rate of change of properties across the space. Thus a response surface approach should yield more accurate estimates.

Extensions beyond the near model constraint. As the KL distance between approximating models and the generating model increases, $-AIC/2n$ becomes an increasingly biased and variable estimate of the Sgf component of the KL distance between the approximating model and the generating model. This effect is strong enough that sometimes very bad models can have low delta AIC values, even sometimes appearing as the best model. It seems reasonable to think that using heteroskedastic nonlinear regressions such as described by Carroll and Ruppert (1988) and Carroll et al. (1995) will allow for incorporating information from much more distant models into the estimated projection. If this does not prove effective, at least the region in which the projection of the generating model resides can be found by plotting the density of AIC good models in the NMDS space. The projection methodology can be applied in the high density region.

As described above, one of the expectations taken in calculating the AIC is over parameter estimates. Estimation of the location and properties of the estimated projection can likely be improved using the reduced variance bias corrected bootstrap information criterion of Kitagawa and Konishi (2010). A benefit of this is that confidence intervals on the estimated projection can be simultaneously calculated. These intervals are based in frequentist probability and can be expressed either as error statistical confidence intervals or as evidential support intervals. This contrasts with intervals produced by model averaging, which despite their sometime presentation as error statistics are actually posterior probability intervals (under a cryptic assumption that the posterior distribution is normal).

Despite the substantial work that still needs to be done, the approach laid out here already has shown a number of clear and substantial advantages over the Bayesian-based model averaging. First, heuristically, the simple ability of being able to visualize model space will aid in the development of new models. Second, the estimated generating model projection is less constrained by the configuration of the model set than is the model average. Third, Sgg, the neg-selfentropy of the generating process itself is estimable. It has long been assumed that this is unknowable, but it can be estimated and is estimated as part of our procedure. In the example it is estimated quite precisely. Sgg as a measure of the dispersion of the generating process is itself of great interest. Fourth, the distance of the generating process from the (hyper)plane of the estimated model space can be estimated. It has long been a complaint of scientific practitioners of model identification through information criteria that they can tell which of their models is closest to the generating process, but they can't tell if any of the models in their model set are any good. Now discrepancy is statistically estimable. Fifth, the strain between a priori and post hoc inference is vacated. The study of the structure of model space corrects for misleading evidence (chance good results), accommodation (over-fitting), and cooking the models. Theoretically, the more models are considered the more robust the scaffolding from which to project the location of the generating process. Currently, the model set for the model projection approach is limited by the near model requirement common to all information criteria analysis. However, as indicated above, non-linear modeling should allow the analyst to bootstrap

(Glymour sense) the structure of model space to validly include in the projection information from models more distant than is valid for model averaging.

Model projection is an evidential alternative to Bayesian model averaging for incorporating information from many models in a single analysis. Model projection, because it more fully utilizes the information in the structure of model space, is able to estimate several very important quantities that are not estimated by model averaging.

References

Akaike, H. (1973). Information theory as an extension of the maximum likelihood principle. In B. N. Petrov & F. Csaki (Eds.), *Second international symposium on information theory* (pp. 267–281). Budapest: Akademiai Kiado.

Akaike, H. (1974). A new look at statistical-model identification. *IEEE Transactions on Automatic Control, AC19*:716–723.

Bandyopadhyay, P., & Brittan, G. (2002). Logical consequence and beyond: A look at model selection in statistics. In J. Wood & B. Hepburn (Eds.), *Logical consequence and beyond*. Oxford: Hermes Science Publishers.

Burnham, K. P., & Anderson, D. R. (2002). *Model selection and multi-model inference: A practical information-theoretic approach* (2nd ed.). New York: Springer.

Burnham, K. P., Anderson, D. R., & Huyvaert, K. P. (2011). AIC model selection and multimodel inference in behavioral ecology: Some background, observations, and comparisons. *Behavioral Ecology and Sociobiology, 65*:23–35.

Carroll, R., & Ruppert, D. (1988). *Transformation and weighting in regression*. London: Chapman and Hall/CRC Reference.

Carroll, R., Ruppert, D., & Stefanski, L. (1995). *Measurement error in non-linear models*. London: Chapman and Hall.

deLeeuw, J. (1992). Introduction to Akaike (1973) information theory and an extension of the maximum likelihood principle. In S. Kotz & N. L. Johnson (Eds.), *Breakthroughs in statistics* (pp. 599–609). London: Springer.

Grace, J. B. (2008). Structural equation modeling for observational studies. *Journal of Wildlife Management, 72*:14–22.

Grace, J. B., & Bollen, K. A. (2006). The interface between theory and data in structural equation models. In USGS. Open-File Report. Available at www.nwrc.usgs.gov/about/web/j_grace.htm.

Grace, J. B., & Keeley, J. E. (2006). A structural equation model analysis of postfire plant diversity in California shrublands. *Ecological Applications, 16*:503–514.

Grace, J. B., & Bollen, K. A. (2008). Representing general theoretical concepts in structural equation models: the role of composite variables. *Environmental and Ecological Statistics, 15*:191–213.

Grace, J. B., Anderson, T. M., Olff, H., & Scheiner, S. M. (2010). On the specification of structural equation models for ecological systems. *Ecological Monographs, 80*:67–87.

Johnson, R. A., & Wichern, W. D. (2002). *Applied multivariate statistical analysis* (6th ed.). Upper Saddle River: Pearson Prentice Hall.

Kitagawa, G., & Konishi, S. (2010). Bias and variance reduction techniques for bootstrap information criteria. *Annals of the Institute of Statistical Mathematics, 62*:209–234.

Taper, M. L., & Ponciano, J. M. (2015). Evidential statistics as a statistical modern synthesis to support 21st century science. *Population Ecology*, doi:10.1007/s10144-015-0533-y.

Afterword

We began the monograph with an African Proverb for working together to go farther. It applies in particular to deepening our understanding of scientific methodology. We end with a quote from Stephen Hawking that highlights our approach in an especially significant way: "[N]ot only does God play dice, but…he sometimes throws them where they cannot be seen." At the level of practice, we believe scientists are bent on finding those hidden dice, whether broken, biased, or unfair. They do so with better and better models to approximate reality, although in the nature of the case those models are strictly false. Methodology is in this general sense probabilistic. But it is more specifically probabilistic as well. Even physics, that traditional paradigm of exactitude, should be construed stochastically with parameters to estimate making room for errors in measurements. The possibility of misleading evidence captures our approach to those model-building at still another level. Statisticians who "get to play in everybody's backyard" help build those models often aimed at estimating parameters as part of probing natural phenomena. At a meta-level, philosophers also have distinctive roles to play. For example, to make scientists and scientifically-minded philosophers aware of their assumptions "where they cannot be seen" especially in their model-building endeavors in which conflating "evidence" and "confirmation" is likely to occur. Developing a credible philosophy of science that escapes this conflation is indispensable. The position we have advanced in this Monograph is based on a multi-disciplinary approach. In our view it is required for a present day multi-disciplinary audience as it has become the "new normal" in both the practice and theory of science.

© The Author(s) 2016
P.S. Bandyopadhyay et al., *Belief, Evidence, and Uncertainty*,
Philosophy of Science, DOI 10.1007/978-3-319-27772-1

Index

© The Author(s) 2016
P.S. Bandyopadhyay et al., *Belief, Evidence, and Uncertainty*,
Philosophy of Science, DOI 10.1007/978-3-319-27772-1